Springer-Verlag Berlin Heidelberg GmbH

Steen Krenk

Mechanics and Analysis of Beams, Columns and Cables

A Modern Introduction to the Classic Theories

Second Edition

With 113 Figures

 Springer

Professor Steen Krenk

Department of Civil Engineering
Technical University of Denmark
Building 118, Brovej
DK-2800 Lyngby / Denmark

Originally published by: Polyteknisk Press, Denmark, 2000

ISBN 978-3-642-62591-6   ISBN 978-3-642-56694-3 (eBook)
DOI 10.1007/978-3-642-56694-3

CIP data applied for

Die Deutsche Bibliothek - CIP-Einheitsaufnahme
Krenk, Steen: Mechanics and analysis of beams, columns and cables /
Steen Krenk. - 2. ed.. - Berlin ; Heidelberg ; New York ; Barcelona ; Hongkong ; London ; Milan ; Paris ;
Singapore ; Tokyo : Springer, 2001
   ISBN 978-3-642-62591-6

Springer-Verlag Berlin Heidelberg New York
a member of BertelsmannSpringer Science+Business Media GmbH

http://www.springer.de
© Springer-Verlag Berlin Heidelberg 2001
Originally published by Springer-Verlag Berlin Heidelberg New York in 2001
Softcover reprint of the hardcover 2nd edition 2001

Typesetting: Camera ready by author
Cover-design: Medio Technologies AG, Berlin
                    SPIN: 10796899   62 / 3020 hu - 5 4 3 2 1 0 -

Preface

The purpose of this book is to illustrate the use of simple mathematical analysis techniques within the area of basic structural mechanics, in particular the elementary theories of beams, columns and cables. The focus is on:

i) Identification of the physical background of the theories and their particular mathematical properties.

ii) Demonstration of mathematical techniques for analysis of simple problems in structural mechanics, and identification of the relevant parameters and properties of the solutions.

iii) Derivation of the solutions to a number of basic problems of structural mechanics in a form suitable for later reference.

The presentation concentrates on the main principles and the characteristics of the solutions. The theory also provides a basis for the formulation of numerical theories and intelligent interpretation of their results.

In this second edition only minor changes have been introduced in the text, and a subject index has been added.

In the preparation of the manuscript I have had valuable assistance from Gunnar Mohr, who read the original manuscript and suggested several improvements of the presentation, and from Norma Hornung, who prepared the illustrations that form an essential part of the book.

Finally, I would like to express my thanks to the COWI Foundation for their financial contribution, and to Ole Jørgensen and Marianne Pedersen of Polyteknisk Forlag, and to the Springer-Verlag for cooperation during the publication process

Birkerød, November 2000 *Steen Krenk*

Contents

Chapter 1

Introduction

Analysis of structural elements like beams, columns and cables is based on the principles of mechanics. The foundation of these principles was laid during the 150 years of the period 1638-1788. In this period the theory of flexible bodies was developed in close interaction with the development of the necessary mathematical techniques, such as differential equations and calculus of variations. Interesting accounts of this development have been given e.g. by Truesdell (1996) and Szabó (1977). Before proceeding to the specific theories of beams, columns and cables in the following chapters we shall briefly summarize some background material on equilibrium of force systems and simple elastic deformation.

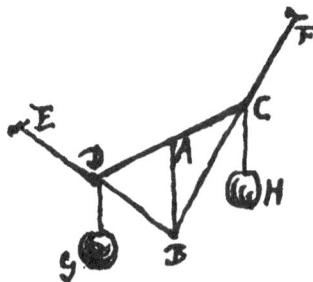

Figure 1.1: Drawing by HUYGENS for addition of forces (1646).

1.1 Force systems

The notion of a force is taken as a basic concept. Forces are represented as vectors, associated with a *point of action*. Thus a force vector $\mathbf{P}$ can be represented in the form $\mathbf{P} = P\mathbf{n}$, where P is the magnitude of the force, and $\mathbf{n}$ is a unit vector indicating the direction of the force. In an orthogonal Cartesian xyz-coordinate system with base vectors $(\mathbf{i}, \mathbf{j}, \mathbf{k})$ a force vector $\mathbf{P}$ is represented by its *components* (P_x, P_y, P_z) according to the relation

$$\mathbf{P} = P_x\mathbf{i} + P_y\mathbf{j} + P_z\mathbf{k} \tag{1.1}$$

Thus the force $\mathbf{P}$ is the vector sum of its components.

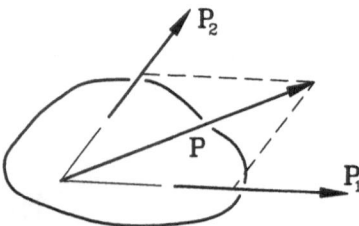

Figure 1.2: Vector addition of two forces through a common point.

A force is associated with a *line of action* passing through the point of action in the direction of the force. Two forces $\mathbf{P}_1$ and $\mathbf{P}_2$ with intersecting lines of action are equivalent to a single force $\mathbf{P}$ obtained as the vector sum, acting at the point of intersection of the two lines of action.

$$\mathbf{P} = \mathbf{P}_1 + \mathbf{P}_2 = (P_{1x} + P_{2x})\mathbf{i} + (P_{1y} + P_{2y})\mathbf{j} + (P_{1z} + P_{2z})\mathbf{k} \tag{1.2}$$

Addition of forces can be done either directly in vector format or in terms of the vector components. However, in order for the force $\mathbf{P}$ to be equivalent to the sum of the two forces $\mathbf{P}_1$ and $\mathbf{P}_2$, their line of action must intersect, thereby defining the line of action of $\mathbf{P}$.

The moment of a force $\mathbf{P}$ about a point O is defined by the vector product

$$\mathbf{M}_O = \mathbf{r} \times \mathbf{P} \tag{1.3}$$

where $\mathbf{r}$ is the vector from the point O to a point on the line of action of the force $\mathbf{P}$. When the point O is not on the line of action the vectors $\mathbf{P}$ and $\mathbf{r}$

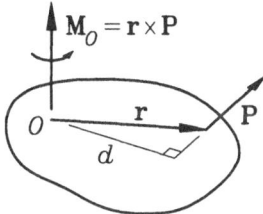

Figure 1.3: Moment $\mathbf{M}_O$ of a force $\mathbf{P}$ about a point O.

define a plane as shown in the figure, and the moment $\mathbf{M}_O$ is normal to that plane. The magnitude of the moment is $|\mathbf{M}_O| = d\,|\mathbf{P}|$, or $M_O = d\,P$, where d is the distance of O from the line of action. The moment of a force can be evaluated as the sum of moments of each of its components.

Example 1.1

Let $\mathbf{P} = (P_x, P_y, P_z)$ be the components in a Cartesian coordinate system of a force acting at a point with coordinates $\mathbf{r} = (x, y, z)$. The use of boldface vector symbols as synonymous with the corresponding Cartesian components is commonly used, although not quite precise. The moment, as defined by the vector product (1.3), is calculated by the formal determinant with the unit base vectors $\mathbf{i}, \mathbf{j}, \mathbf{k}$ in the first row,

$$\mathbf{M}_O = \mathbf{r} \times \mathbf{P} = \begin{vmatrix} \mathbf{i} & \mathbf{j} & \mathbf{k} \\ x & y & z \\ P_x & P_y & P_z \end{vmatrix} = \left\{ \begin{array}{l} (yP_z - zP_y)\,\mathbf{i} \\ + (zP_x - xP_z)\,\mathbf{j} \\ + (xP_y - yP_x)\,\mathbf{k} \end{array} \right.$$

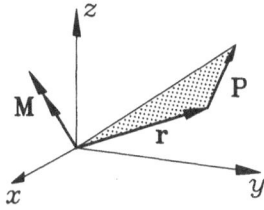

Figure 1.4: Moment $\mathbf{M}_O$ of $\mathbf{P}$ about O. Normal to the plane containing vectors $\mathbf{r}$ and $\mathbf{P}$.

The moment vector $\mathbf{M}_O$ is normal to the plane containing the vectors $\mathbf{r}$ and $\mathbf{P}$ as shown in Fig. 1.4. The symbol with double arrowhead is used to denote moment vectors.

In the particular case of a plane problem in the xy-plane the force and position vectors will have the components (P_x, P_y) and (x, y), while $P_z = 0$ and $z = 0$. In this case the moment about O is given by the component

$$M_z = xP_y - yP_x$$

while $M_x = M_y = 0$. In two-dimensional problems it is customary to define a positive direction of rotation in the plane, and use this definition to define positive moments.

Two forces of equal magnitude and opposite direction, acting along two different lines are called a *force couple*. The vector sum of a force couple is zero, but because the lines of action are different the two forces do not cancel. Figure 1.5 shows the plane containing the two lines of action. For any point O the moment of the two forces is

$$
\begin{aligned}
\mathbf{M}_O &= \mathbf{r}_1 \times \mathbf{P} + \mathbf{r}_2 \times (-\mathbf{P}) \\
&= (\mathbf{r}_1 - \mathbf{r}_2) \times \mathbf{P} = \mathbf{d} \times \mathbf{P}
\end{aligned}
\tag{1.4}
$$

where $\mathbf{d}$ is a vector connecting the two lines of action. Thus, the moment is independent of the location of the point O, and the force couple is equivalent to the moment $\mathbf{d} \times \mathbf{P}$.

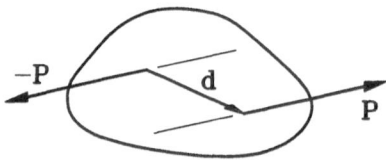

Figure 1.5: Force couple, two opposite forces with distance d.

A force $\mathbf{P}$ can be translated the distance $\mathbf{d}$ when a moment $\mathbf{M} = -\mathbf{d} \times \mathbf{P}$ is added. The procedure consists in adding the zero force $\mathbf{P} - \mathbf{P}$ at the new position, and then combining the original force with $-\mathbf{P}$ into a force couple as shown in the Fig. 1.6.

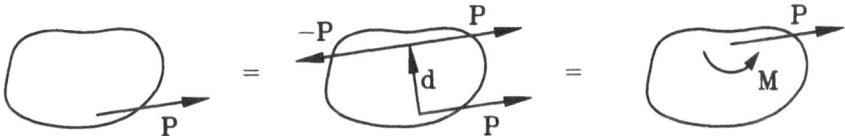

Figure 1.6: Translating a force by addition of a moment.

It follows from the equivalence between force pairs and moments, that it is natural to include moments in the concept of a force system. Two force systems are said to be equivalent, if they have the same sum of the individual forces, $\mathbf{P} = \Sigma_i \mathbf{P}_i^a = \Sigma_j \mathbf{P}_j^b$, and their total moment about an arbitrary point are identical, $\mathbf{M} = \Sigma_k \mathbf{M}_k^a + \Sigma_i \mathbf{r}_i^a \times \mathbf{P}_i^a = \Sigma_l \mathbf{M}_l^b + \Sigma_j \mathbf{r}_j^b \times \mathbf{P}_j^b$. The translation of a force, shown in Fig. 1.6 is a special instance of two equivalent force systems.

1.2 Equilibrium

Statics is the science of particles and bodies that remain at rest under the action of systems of forces and moments in equilibrium . A system of forces and moments is in equilibrium when the vector sum of all the forces *and* the sum of all applied moments plus the moments of all forces about an arbitrary point vanish.

$$\mathbf{P} = \Sigma_i \mathbf{P}_i = 0 \quad , \quad \mathbf{M}_O = \Sigma_j \mathbf{M}_j + \Sigma_i \mathbf{r}_i \times \mathbf{P}_i = 0 \qquad (1.5)$$

In three-dimensional problems equilibrium amounts to six scalar equations, while in two-dimensional problems there are only two force components and one moment component, leading to three equilibrium equations.

In some instances it is convenient to use moments about different points to express equilibrium. In two-dimensional problems equilibrium may be expressed by use of one, two or three moment conditions as shown below. When using moments about two points A and B the direction of projection must not be orthogonal to the line joining A and B, and when using moments

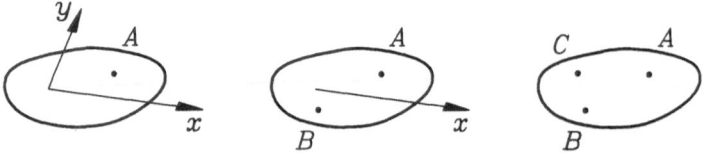

Figure 1.7: Equilibrium of a plane force system by a
total of three projections and moments.

about three points A, B and C they must not lie on a line.

$$\left. \begin{array}{rcl} \Sigma_i P_{ix} & = & 0 \\ \Sigma_i P_{iy} & = & 0 \\ \Sigma_j M_j^A & = & 0 \end{array} \right\} \text{ or } \left\{ \begin{array}{rcl} \Sigma_i P_{ix} & = & 0 \\ \Sigma_j M_j^A & = & 0 \\ \Sigma_j M_j^B & = & 0 \end{array} \right\} \text{ or } \left\{ \begin{array}{rcl} \Sigma_j M_j^A & = & 0 \\ \Sigma_j M_j^B & = & 0 \\ \Sigma_j M_j^C & = & 0 \end{array} \right. \quad (1.6)$$

The moment conditions contain the moments of all forces as well as any moments contained in the system, as indicated by using a different summation
subscript in (1.6). Similar alternative sets of equilibrium conditions can be
formulated for three-dimensional problems, but will not be used here.

1.3   Supports and reactions

In structural mechanics it is common to consider problems where the structure or structural element is acted on by a non-equilibrium force system,
often called the load. The load is carried by *reactions* provided via support
of the structure. Figure 1.8 illustrates three common types of support for
an end of a beam. In the rigid support of Fig. 1.8a the end is constrained
in such a way that displacement $\mathbf{u}$ and rotation $\boldsymbol{\theta}$ is prevented. In order to
prevent displacement the support must be able to provide the necessary force
$\mathbf{P}$, and in order to prevent rotation the support must be able to provide the
necessary moment $\mathbf{M}$. The difference between loads and reactions is that
while the loads are assumed to be known in the analysis, the reactions have
to be determined by the analysis. In the rigid support the constraints prevent both displacement and rotation, and at this end the beam can therefore
support both a reaction force and a reaction moment. In the simple support
of Fig. 1.8b the beam end is free to rotate, and it can therefore not support

a reaction moment. In the free end of Fig. 1.8c neither reaction force or reaction moment can be supported.

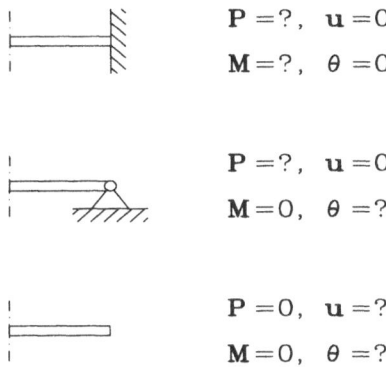

$$\mathbf{P} = ?, \quad \mathbf{u} = 0$$
$$\mathbf{M} = ?, \quad \theta = 0$$

$$\mathbf{P} = ?, \quad \mathbf{u} = 0$$
$$\mathbf{M} = 0, \quad \theta = ?$$

$$\mathbf{P} = 0, \quad \mathbf{u} = ?$$
$$\mathbf{M} = 0, \quad \theta = ?$$

Figure 1.8: Typical beam support conditions: a) rigid support, b) simple support, c) free end.

It is seen that there is a duality between the force and displacement vectors and between the moment and rotation vectors. Either one of each pair can be prescribed. This duality holds for each component of the vectors. Thus either the displacement component u_x or the corresponding force component P_x can be prescribed, and similarly for the other components. In general this permits a large number of combinations, most conveniently treated in connection with specific problems. This duality between *static variables* $\mathbf{P}, \mathbf{M}$ and the corresponding *kinematic variables* $\mathbf{u}, \theta$ is closely connected with the principle of virtual work, discussed at several places below.

In a static analysis of a structure or structural element the first task is often to determine the reactions corresponding to the support conditions of the structure. The following example illustrates the alternative equilibrium conditions from Fig. 1.7.

Example 1.2

Figure 1.9 shows a two-dimensional beam problem. The beam of length ℓ extends from point A to point B. At point A the beam has a fixed simple support, permitting a horizontal reaction force component R_H and a vertical reaction force component R_A. The simple support at point B is free to move horizontally, and therefore supports only a vertical reaction force component R_B.

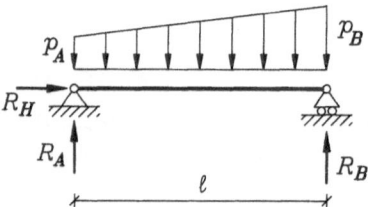

Figure 1.9: Simply supported beam with distributed load
$p(x)$ with linear variation between p_A and p_B.

The beam carries a distributed vertical load of intensity $p(x)$ per unit length with linear variation between p_A and p_B. Introducing a length coordinate x from the left end point A then gives the load as

$$p(x) \; = \; p_A \left(1 - \frac{x}{\ell} \right) \; + \; p_B \, \frac{x}{\ell}$$

This load is equivalent to a vertical force of magnitude

$$P \; = \; \int_0^\ell p(x) \, \mathrm{d}x \; = \; \tfrac{1}{2}(p_A + p_B)\ell$$

Equilibrium in horizontal and vertical directions immediately gives the following two relations for the reactions

$$R_H \; = \; 0 \quad , \qquad R_A + R_B \; - \; P \; = \; 0$$

The first relation gives the horizontal reaction explicitly, while the second gives a relation between the vertical reactions R_A and R_B. The remaining relation is conveniently chosen as the moment about A,

$$M_A \; = \; R_B \, \ell \; - \; \int_0^\ell p(x) \, x \, \mathrm{d}x \; = \; 0$$

whereby

$$R_B \; = \; \frac{1}{\ell} \int_0^\ell p(x) \, x \, \mathrm{d}x \; = \; \tfrac{1}{6}p_A\ell \; + \; \tfrac{1}{3}p_B\ell$$

The vertical reaction has now been determined explicitly, and the reaction R_A follows from the vertical projection as

$$R_A = P - R_B = \tfrac{1}{3}p_A\ell + \tfrac{1}{6}p_B\ell$$

This is the procedure of Fig. 1.7a, using two projections and one moment equation.

In the present case it would have been equally simple to use horizontal projection in combination with moments about the end points A and B, according to the procedure of Fig. 1.7b. The moment about A gives R_B explicitly as above, and moment about B gives R_A by symmetry by simply interchanging the role of p_A and p_B. Many problems of structural mechanics possess some kind of symmetry, that can be used in the solution process or to check the consistency of the results.

In the example the reactions could be determined completely from equilibrium considerations. For two-dimensional problems there are three equilibrium conditions, e.g. in the form of two projections and a moment equation. Thus, the equilibrium conditions for a single body in two dimensions give three equations for the components of the reactions. If the support conditions permit complete determination of the reactions from the equilibrium conditions the problem is called *statically determinate*. If the equilibrium conditions permit only incomplete determination of the reactions, the problem is called *statically indeterminate*. The beam of Example 1.2 would become statically indeterminate, if e.g. both ends were rigidly fixed. Each end would then permit two reaction force components and a moment, making a total of six reaction components. For statically indeterminate problems the equilibrium equations only determine a system of reactions that is equivalent to the actual reactions, while the detailed distribution of the load between the reactions depends on the properties of the loaded body, e.g. its elastic deformability. Thus, a full analysis of statically indeterminate structures requires static equilibrium as well as a description of the deformations of the structure under the prescribed load. Statically indeterminate problems of elastic beams are treated in the next chapter.

1.4   Internal forces

A full static analysis of a structure or structural element includes not only determination of the reactions but also a description of how the forces and moments are transferred through the structure. The present text is confined to one-dimensional structural elements like beams, columns and cables, and the principles of the so-called *internal forces* can therefore be described in elementary terms. The basic idea is to imagine that a section of the structure is cut through. For a simple structure like a beam this will separate the structure into two parts. This is illustrated in Fig. 1.10 for a section located at arc length $s = s_0$ along the beam. The left part of the beam, corresponding to $s < s_0$ is acted on by a force $\mathbf{N}(s_0)$ and a moment $\mathbf{M}(s_0)$. By the rule of action and reaction the right side of the beam, corresponding to $s > s_0$, is acted on by the opposite force and moment $-\mathbf{N}(s_0)$ and $-\mathbf{M}(s_0)$. The force and moment $\mathbf{N}$ and $\mathbf{M}$ are called the internal or generalized forces at s_0. The moment $\mathbf{M}(s_0)$ is traditionally included in the term internal forces. In standard usage the force and moment vectors acting on the beam part with $s < s_0$ are denoted $\mathbf{N}(s_0)$ and $\mathbf{M}(s_0)$, while the force and moment on the other part must then include a minus sign, i.e. $-\mathbf{N}(s_0)$ and $-\mathbf{M}(s_0)$.

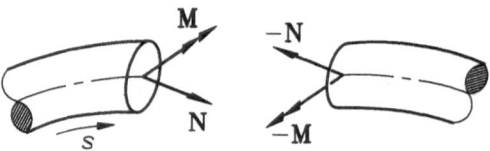

Figure 1.10: Internal forces $\mathbf{N}(s_0)$ and $\mathbf{M}(s_0)$ at s_0.

For each part of the beam the internal force $\mathbf{N}(s_0)$ and the moment $\mathbf{M}(s_0)$ are initially unknown, and therefore to be considered like reactions. For a statically determinate structure the internal forces can be determined from the equilibrium equations alone, while the internal forces of a statically indeterminate structure depends on the deformation properties of the structure. The following example is a continuation of the two-dimensional beam problem from Example 1.2.

Example 1.3

Figure 1.11 shows the two-dimensional beam problem from Example 1.2. The length of the beam is measured by the coordinate x, and the beam is separated into two parts by a section at the distance x from the left end point A. The force vector on the section is represented by the axial component $N(x)$, called the normal force, and the transverse component $Q(x)$, called the shear force. The moment $M(x)$ is chosen as positive for compression in the top of the beam.

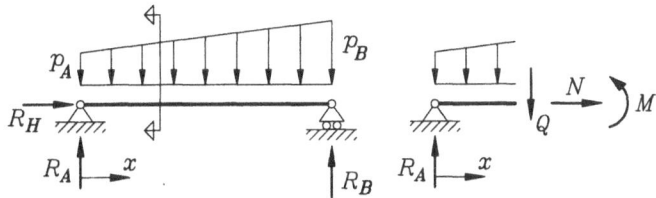

Figure 1.11: Internal forces $N(x)$, $Q(x)$ and $M(x)$ in continuously loaded beam.

The beam is statically determinate, and the internal forces can therefore be determined from the equilibrium conditions alone. The normal force follows directly from a horizontal projection,

$$N(x) \;=\; -R_H \;=\; 0$$

The shear force follows from a vertical projection,

$$
\begin{aligned}
Q(x) \;&=\; R_A \;-\; \int_0^x p(s)\,ds \;=\; R_A \;-\; \tfrac{1}{2}\big(p_A + p(x)\big)\,x \\
&=\; \tfrac{1}{3}p_A\ell + \tfrac{1}{6}p_B\ell \;-\; \tfrac{1}{2}\Big(p_A + p_A\big(1 - \tfrac{x}{\ell}\big) + p_B\,\tfrac{x}{\ell}\Big)\,x \\
&=\; \frac{p_A}{6\ell}(2\ell^2 - 6\ell x + 3x^2) \;+\; \frac{p_B}{6\ell}(\ell^2 - 3x^2)
\end{aligned}
$$

Clearly the problem is symmetric, if the figure is turned left to right. Symmetry would imply that the p_A and p_B terms are similar when x is replaced by $x' = \ell - x$. Indeed, if x is replaced in the first term by this substitution, we obtain a fully symmetric form of the shear force,

$$Q(x) \;=\; -\frac{p_A}{6\ell}(\ell^2 - 3x'^2) \;+\; \frac{p_B}{6\ell}(\ell^2 - 3x^2)$$

Each of the two terms correspond to the shear force from a triangular load distribution. Note that $Q(0) = R_A$ and $Q(\ell) = -R_B$, corresponding to the sign convention adopted here.

The internal moment $M(x)$ is calculated by taking the moment about the point of the section at x.

$$
\begin{aligned}
M(x) &= R_A x - \int_0^x p(s)(x-s)\,ds \\
&= R_A x - (\tfrac{1}{2}p_A x)\tfrac{2}{3}x - (\tfrac{1}{2}p(x)x)\tfrac{1}{3}x
\end{aligned}
$$

The integral has been calculated by considering the load as two triangular parts with maximum load intensity p_A and $p(x)$, respectively. The center of a triangular load distribution is located $\tfrac{1}{3}x$ from the end with maximum load. The reaction R_A and the load distribution function $p(x)$ are substituted from Example 1.2, whereby

$$
M(x) = \frac{p_A}{6\ell}(2\ell^2 x - 3\ell x^2 + x^3) + \frac{p_B}{6\ell}(\ell^2 x - x^3)
$$

As in the case of the shear force this solution does not directly demonstrate the symmetry of the problem with respect to interchange of left and right, but after introduction of the variable $x' = \ell - x$ the internal moment takes the symmetric form

$$
M(x) = \frac{p_A}{6\ell}(\ell^2 - x'^2)x' + \frac{p_B}{6\ell}(\ell^2 - x^2)x
$$

Note, that the moment vanishes at the ends of the beam, $M(0) = M(\ell) = 0$, as prescribed by the boundary conditions.

It is interesting to note that the shear force and internal moment satisfy the differential relations

$$
\frac{dQ(x)}{dx} = -p(x) \quad , \quad \frac{dM(x)}{dx} = Q(x)
$$

This suggests an alternative solution procedure, based on differential equations and interpretation of the support conditions as boundary conditions on the functions $Q(x)$ and $M(x)$. This approach is developed in Chapter 2.

———————

1.5   Principle of virtual work

The equilibrium equations (1.5) for forces $\mathbf{P}_i$ and moments $\mathbf{M}_j$ acting on a rigid body can be combined into a statement of *virtual work*. For a set of forces $\mathbf{P}_i$ equilibrium requires that the projection of the sum of the forces $\sum_i \mathbf{P}_i$ on an arbitrary direction must vanish. This statement can be viewed alternatively as a statement that the virtual work $\delta\mathbf{u} \cdot \sum_i \mathbf{P}_i$, performed when the forces are moved as if they were acting on a rigid body subjected to an arbitrary *virtual displacement* $\delta\mathbf{u}$, must vanish. This is illustrated in Fig. 1.12a.

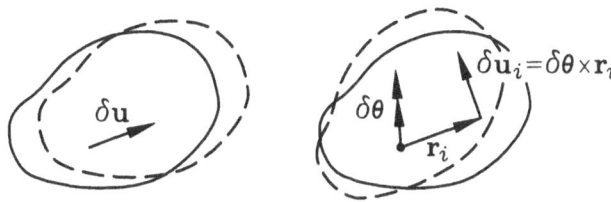

Figure 1.12: Virtual displacements due to: a) Rigid body translation $\delta\mathbf{u}$, b) Rigid body rotation $\delta\boldsymbol{\theta}$.

The problem can be generalized by considering also a virtual infinitesimal rotation $\delta\boldsymbol{\theta}$ of the rigid body about the point O. At a point with position vector $\mathbf{r}_i$ relative to O the virtual infinitesimal rotation $\delta\boldsymbol{\theta}$ gives the displacement $\delta\boldsymbol{\theta} \times \mathbf{r}_i$ as shown in Fig. 1.12b. Thus, a general virtual displacement consisting of an infinitesimal rotation $\delta\boldsymbol{\theta}$ about O plus a translation $\delta\mathbf{u}$, is expressed in the form

$$\delta\mathbf{u}_i = \delta\mathbf{u} + \delta\boldsymbol{\theta} \times \mathbf{r}_i \tag{1.7}$$

The contribution of a force $\mathbf{P}_i$ at $\mathbf{r}_i$ to the virtual work is $\delta\mathbf{u}_i \cdot \mathbf{P}_i$, where $\delta\mathbf{u}_i$ is the virtual displacement at $\mathbf{r}_i$, while the contribution from an applied moment $\mathbf{M}_j$ is $\delta\boldsymbol{\theta} \cdot \mathbf{M}_j$, with the common rotation $\delta\boldsymbol{\theta}$ for all moments. Thus, the total virtual work of the external forces and moments is

$$\begin{aligned}
\delta W_{ex} &= \sum_i \delta\mathbf{u}_i \cdot \mathbf{P}_i + \sum_j \delta\boldsymbol{\theta} \cdot \mathbf{M}_j \\
&= \delta\mathbf{u} \cdot \sum_i \mathbf{P}_i + \delta\boldsymbol{\theta} \cdot \left(\sum_i \mathbf{r}_i \times \mathbf{P}_i + \sum_j \mathbf{M}_j \right)
\end{aligned} \tag{1.8}$$

where the last result is obtained by using that the dot and cross product signs can be interchanged, according to the vector identity

$$\delta\boldsymbol{\theta}\times\mathbf{r}_i\cdot\mathbf{P}_i \;=\; \delta\boldsymbol{\theta}\cdot\mathbf{r}_i\times\mathbf{P}_i \tag{1.9}$$

According to the conditions (1.5) equilibrium means that the sum of the external forces and moments

$$\mathbf{P}_{ex} \;=\; \textstyle\sum_i \mathbf{P}_i \qquad,\qquad \mathbf{M}_{ex} \;=\; \textstyle\sum_i \mathbf{r}_i\times\mathbf{P}_i + \textstyle\sum_j \mathbf{M}_j \tag{1.10}$$

must vanish. Thus, for a set of forces and moments in equilibrium the virtual work corresponding to any *arbitrary* virtual translation $\delta\mathbf{u}$ and virtual infinitesimal rotation $\delta\boldsymbol{\theta}$ must vanish.

$$\delta W_{ex} \;=\; \delta\mathbf{u}\cdot\mathbf{P}_{ex} + \delta\boldsymbol{\theta}\cdot\mathbf{M}_{ex} \;=\; 0 \tag{1.11}$$

This is the equation of virtual work for a virtual displacement field (1.7) corresponding to a rigid body.

It is important to understand the relation between the force and moment vector equilibrium equations (1.5) and the corresponding statement of vanishing virtual work in (1.11). The equilibrium conditions (1.5) consist of two vector equations, one for vanishing total force and one for vanishing total moment. If expressed in vector components this amounts to six scalar equations. Apparently the virtual work equation (1.11) corresponds to only a single scalar equation. However, the equation of vanishing virtual work is valid for *any* choice of the virtual vectors $\delta\mathbf{u}$ and $\delta\boldsymbol{\theta}$. Thus, different choices of these vectors produce different equations. As six virtual vector components can be chosen independently, six independent equilibrium conditions can be produced.

Example 1.4

In Example 1.2 the reactions of the simply supported beam in Fig. 1.9 were determined by use of the equilibrium equations. Alternatively the reactions can be determined by use of the principle of virtual work. The horizontal reaction would follow from a horizontal translation of the beam and its load and reactions. The vertical reactions are determined by virtual rotations about the end points of the beam.

Figure 1.13 shows the determination of the reaction R_B by a virtual rotation of the beam about the end point A. Note, that the choice of the point A as the center of rotation implies that the reactions do not contribute to the virtual work in this case. If the downward displacement of point B is selected as one unit, the virtual displacement field is

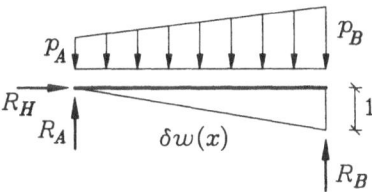

Figure 1.13: Virtual displacement $\delta w(x)$ for determination
of the reaction R_B.

$\delta w(x) = x/\ell$. There are no moments, and the virtual work equation is
written in terms of the forces and their virtual translations,

$$\delta W_{ex} = 1 \cdot (-R_B) + \int_0^\ell \delta w(x)\, p(x)\, \mathrm{d}x = 0$$

Substitution of the virtual displacement field $\delta w(x) = x/\ell$ then gives
the reaction as

$$R_B = \int_0^\ell \delta w(x)\, p(x)\, \mathrm{d}x = \frac{1}{\ell}\int_0^\ell x\, p(x)\, \mathrm{d}x$$

This is seen to be equivalent to the moment equation used in Example
1.2. The integral is most conveniently computed by considering the load
as consisting of two triangles, whereby

$$R_B = \tfrac{1}{6}p_A\ell + \tfrac{1}{3}p_B\ell$$

In this simple case the principle of virtual work appears as a geometric
illustration of moment equilibrium, analogous to a lever where the load
contributes in proportion to its distance from the center of rotation.

In Example 1.4 the correspondence between the direct definition of moments
and the virtual work due to infinitesimal rotations was so close that the
principle of virtual work seemed nearly trivial. The usefulness of the principle
of virtual work in the determination of reactions in statics is perhaps better
appreciated, when used to structures containing parts connected by hinges.
In that case different virtual rigid body motions may be used for the parts
connected by the hinge. The force transferred through the hinge will not

contribute to the virtual work, if the two parts remain connected by the hinge, and the moment vanishes at the hinge. The general approach is illustrated by the following simple example.

Example 1.5

Figure 1.14 shows a system of two beams connected by a hinge at point D. Both beams carry a vertical load of constant intensity p over their full length.

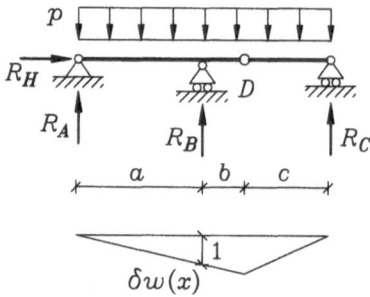

Figure 1.14: Virtual displacement $\delta w(x)$ for determination of reaction R_B.

Each reaction can be determined explicitly, without involving the other reactions, by the principle of virtual work. The idea is to devise a rigid body displacement field for each of the two beams such that they remain connected at the hinge. If only one of the reactions is displaced by the virtual displacement field, an explicit equation for this reaction is obtained.

The figure shows the virtual displacement field for the reaction R_B. The support at B is lowered one unit. The beam AD rotates as a rigid body about A corresponding to the displacement field $\delta w(x) = x/a$ *for this beam*. The beam DC rotates about C such that the two beams remain connected via the hinge at D. The displacement at D is determined by the rotation of the beam AD, from which

$$\delta w_D \;=\; \delta w(a{+}b) \;=\; \frac{a+b}{a}$$

This determines the triangular displacement $\delta w(x)$ for both beams as shown in the figure.

The equation of virtual work now includes the sum of the virtual work
of both the beams. They interact by a shear force at the hinge, but due
to the identical displacement of the two opposite shear forces there is no
contribution to the virtual work from the shear force.

$$\delta W_{ex} \;=\; 1\cdot(-R_B) \;+\; \int_A^C \delta w(x)\, p\, \mathrm{d}x \;=\; 0$$

This equation determines the reaction R_B as

$$R_B \;=\; \int_A^C \delta w(x)\, p\, \mathrm{d}x \;=\; \tfrac{1}{2}\delta w_D\, p(a+b+c) \;=\; \frac{(a+b)(a+b+c)}{2\,a}\, p$$

where the integral of the virtual displacement $\delta w(x)$ follows immediately
from its triangular form in Fig. 1.14.

The theory and examples of this section were concerned with the principle of
virtual work for a virtual displacement field of the form (1.7) corresponding
to moving forces and moments in accordance with one or more rigid bodies.
It will be demonstrated in Chapter 2 that the principle of virtual work can
be extended to flexible bodies. Just like the virtual work of rigid bodies
establishes a connection between the form of the equilibrium equations (1.5)
and the representation of the displacements of a rigid body (1.7), the more
general principle establishes a relation between equilibrium and deformation
of flexible bodies. The principle of virtual work is not only a convenient
instrument for specific computations, but also an important property of the
mathematical equations that describe the mechanics of the problem.

1.6   Stress, strain and elasticity

Structures generally deform under load, although the deformation is usually
too small to be seen directly. In spite of its small magnitude the deformation
may be important, e.g. for the detailed distribution of the reactions for
statically indeterminate structures and for the stability of columns.

Some of the basic concepts are conveniently introduced by reference to a
homogeneous bar, which in its unloaded state has length ℓ and cross-section
area A. The bar is loaded in tension by an axial force P as shown in Fig. 1.15.
It follows from the equilibrium conditions that the internal axial force at each

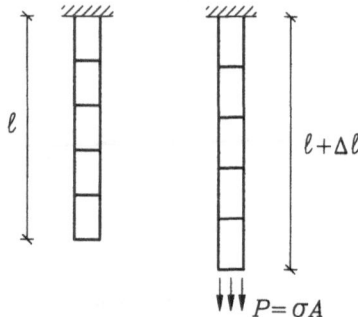

Figure 1.15: Homogeneous bar of initial length ℓ and cross-section area A.

cross-section is $N = P$. Any part of the bar is in the same state of loading. This state is characterized by normalizing the force by the cross-section area to define the *stress*

$$\sigma \;=\; \frac{N}{A} \tag{1.12}$$

The force N and the stress σ act normal to the cross section, and σ is therefore called the *normal stress*.

The deformation of the bar is described by its elongation $\Delta\ell$. The initial length in the unloaded state is ℓ and after loading by the force P the length is $\ell + \Delta\ell$. The elongation is uniformly distributed over the length as indicated in Fig. 1.15, and it is therefore convenient to introduce the normalized elongation

$$\varepsilon \;=\; \frac{\Delta\ell}{\ell} \tag{1.13}$$

The non-dimensional measure of deformation ε is called the normal strain.

Figure 1.16 illustrates the the concepts of normal stress and normal strain with reference to a cube of unit side length. The top and bottom faces of the cube are acted on by vertical forces of magnitude σ in the outward direction, corresponding to tension in the cube. The corresponding normal strain ε is shown as an elongation of the cube in the vertical direction.

The magnitude of the loading of the material in the bar has now been described by the stress σ and its deformation by the strain ε. The simplest assumption of material behaviour connecting stress and strain is a linear

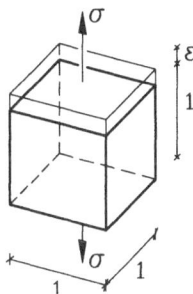

Figure 1.16: Unit cube with normal stress σ and corresponding normal strain ε.

relation of the form

$$\sigma = E\varepsilon \qquad (1.14)$$

This type of material behaviour is called linear elastic. The material parameter E is called the *modulus of elasticity*. It represents the stiffness of the material – the higher the modulus of elasticity, the smaller the deformation. Like stress its dimension is force per area. If the force is measured in Newton and the area in square meter, the modulus of elasticity has the unit Pascal, denoted Pa $= N/m^2$. Some typical values of the modulus of elasticity are listed in Table 1.1.

Table 1.1: Modulus of elasticity.

Material	MPa
Rubber	7
Nylon	1400
Plywood	7000
Wood	14000
Concrete	30000
Aluminum	70000
Steel	210000

The elastic relation (1.14) was first proposed by ROBERT HOOKE 1675 in the form of an anagram with the solution *ut tensio sic vis*, meaning *as the extension so the force*. Hooke verified the relation by a series of experiments published in 1678, see Truesdell (1960) pp. 53. The relation (1.14) is often called Hooke's law. It is based on the assumption that σ is the only stress component in the material, corresponding to vertical load of the cube in Fig. 1.16. For most elastic materials this *uniaxial load* will produce the extension ε as predicted by (1.14) and in addition a contraction transverse to the direction of the load. The effect of transverse contraction is usually not included in beam theories, and will not be treated here. A description of transverse contraction may be found in any textbook on mechanics of materials, e.g. Gere & Timoshenko (1997).

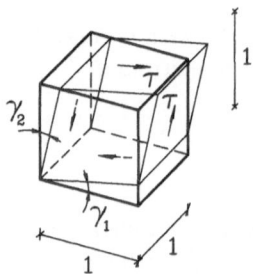

Figure 1.17: Unit cube with shear stress τ and corresponding shear strain γ.

Although tension and compression are the most important effects in beams, shear deformation may also occur. Shear deformation can be illustrated by placing a telephone book on a desk and pushing the top cover towards the front of the book. All pages retain their original size but slide a little bit in the direction of the push. The basic shear mechanism is illustrated in Fig. 1.17 for a cube of unit side length. The top and bottom faces are loaded by forces of magnitude τ in their own plane towards the right and the left, respectively. Equilibrium requires that the left and right faces are loaded with a downward and an upward force of the same magnitude to prevent rotation of the cube. This is the basic shear load. The shear load will rotate the sides of the cube as shown in the figure as γ_1 and γ_2. The change of the angle between the faces of the cube is the *shear strain*

$$\gamma = \gamma_1 + \gamma_2 \qquad (1.15)$$

The shear strain is a change of angle between the faces, and is independent of any overall rotation of the cube.

In a linear elastic material the shear stress τ is proportional to the shear strain γ,

$$\tau = G\gamma \qquad (1.16)$$

The coefficient G is a material parameter, called the *shear!modulus*. It is similar to the modulus of elasticity and also has the dimension of force per area. For an isotropic material – that is a material in which the properties are independent of the direction in the material – there is a connection between the modulus of elasticity E and the shear modulus G. The relation implies that

$$\tfrac{1}{3}E \ < \ G \ < \ \tfrac{1}{2}E \tag{1.17}$$

A typical value is $G \simeq \tfrac{3}{8}E$. Further details may be found e.g. in Gere & Timoshenko (1997).

1.7 Summary

This chapter has given a brief summary of the addition of forces as vectors and the equilibrium conditions for a system of forces and moments. The general equations are given in vector form (1.5), and different specialized forms for two-dimensional problems are given in (1.6) and illustrated in Fig. 1.7.

Reactions are defined as forces and moments generated by the support conditions, typically unknown at the beginning of the analysis. A duality between prescribing *either* a force/moment component *or* the corresponding displacement/rotation component was discussed and illustrated in Fig. 1.8. This theme is explored further in Chapters 2 and 3 in connection with differential equations and boundary conditions for beams and columns.

The concept of internal forces was introduced as the forces and moments exchanged across an imaginary section through the structure, Fig. 1.10. The imaginary section separates the structure in two parts, and a force and a moment is needed to preserve the state of each part unchanged. Example 1.3 illustrated the computation of internal forces in a simple beam, and also demonstrated that the internal force and moment functions satisfy differential equations.

The principle of virtual work was established for a displacement field corresponding to a rigid body. It is based on the similarity between the vector equilibrium conditions (1.5) and the displacement field (1.7) corresponding to a virtual translation and rotation. The result simply says that the total work performed by *all* external forces and moments, including reactions, must vanish for an *arbitrary* virtual rigid body displacement field. The extension of the principle to more than one rigid body displacement field for a structure with a hinge was illustrated in Example 1.5. Chapter 2 will generalize the principle of virtual work to flexible beams.

Finally the concept of homogeneous deformation in the form of *extension* and *shear* was described, and particular cases of Hooke's law for extension and shear were introduced. These will be used to describe the bending and shear deformation of beams in Chapter 2.

1.8 Exercises

Exercise 1.1 Make a sketch that illustrates the moment vector from one component P_x of a force (P_x, P_y, P_z) acting at the point with coordinates (x, y, z) about the origin $(0, 0, 0)$. (It is a good idea to plot the components of the moment about the coordinate axes)

Exercise 1.2 Determine the line of action of the force $\mathbf{P}$ that is equivalent to the distributed load in Example 1.2.

Exercise 1.3 Determine all the reactions on the beam system from Example 1.5 shown in Fig. 1.14. Find the shear force at the hinge, and indicate its direction on a figure. Find the internal moment at the support B.

Exercise 1.4 Replace the load on the beam system shown in Fig. 1.14 with a downward concentrated force P at point D, and find all reactions by the principle of virtual work. Also determine the shear force Q_{D-} at a point just left of D and the internal moment M_B at B. Plot the moment distribution $M(x)$ along both beams.

Chapter 2

Beams

The analysis of beams is usually dated from the treatise 'Two New Sciences' by GALILEO GALILEI (1638). Galilei treated questions like the relative strength of beams with rectangular cross-section and arrived at correct scaling laws for dependence on width, height and length. He also attempted an analytical description of the bending of an elastic beam. However the analysis was incomplete due to an error regarding the size of the tension and compression zones in the beam and lack of sufficiently general analytical tools, like differential calculus.

The basic mechanism in bending of elastic beams was identified by ROBERT HOOKE in 1678 and is essentially that described in the following section. Hooke arrived at the mechanism without need of sophisticated mathematical analysis by considering a simple state of pure bending, while Gallilei had attempted the solution of the more general problem of a cantilever beam loaded by a transverse end force. This problem was treated several times with general assumptions of large deformations by JAMES BERNOULLI in 1694 and the following years. Bernoulli assumed that cross-sections that were initially plane would remain plane after deformation and that these plane sections would remain orthogonal to the beam axis. Beam theories based on these assumptions are often termed 'Bernoulli Beam Theory'. As seen from Fig. 2.1 the analysis was by no means simple, and the two small triangles within the beam indicate that Bernoulli implicitly assumed that the beam cross-section rotated around the compression side of the beam. Although this is not correct, Bernoulli's analyses mark an important step in the formulation of theories for flexible bodies.

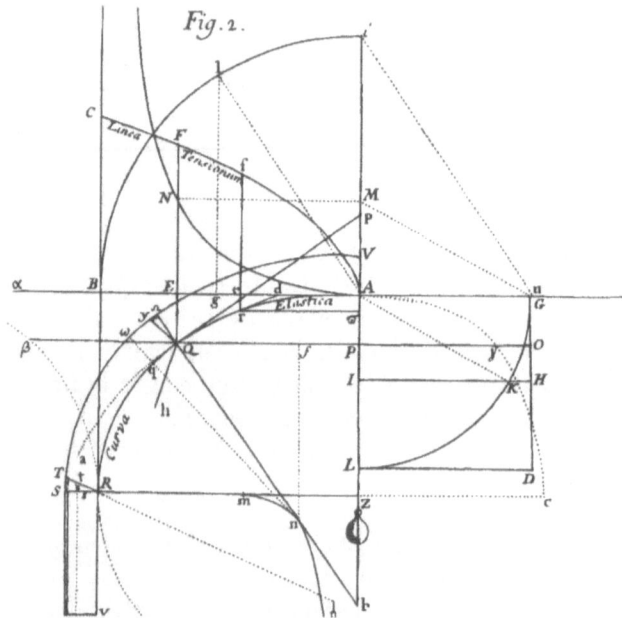

Figure 2.1: JAMES BERNOULLI's first publication of the elastica (1694).

2.1   The mechanics of beam bending

In most cases the bending deformations constitute the dominating contribution to the total deformation of a beam, and many beam problems can be
investigated by considering only the bending deformation. The bending deformation mechanism is identified by considering bending of a straight beam
with constant cross-section by equal but opposite end moments as shown in
Fig. 2.2. The end moments M establish a state of constant internal moment
$M(s) \equiv M$ at all cross-sections of the beam, and this enables a very simple
description of the state of deformation. The beam cross-section is assumed
to be symmetric with respect to the plane of the paper, and by symmetry
the deformed beam will remain in the original plane.

Now, consider the cross-section AB. In the initial undeformed state, shown
in the lower part of Fig. 2.2 the cross-section AB is plane, and orthogonal
to the beam axis. The deformed beam is symmetric about AB, and thus
the cross-section AB remains plane and orthogonal to the beam axis in the

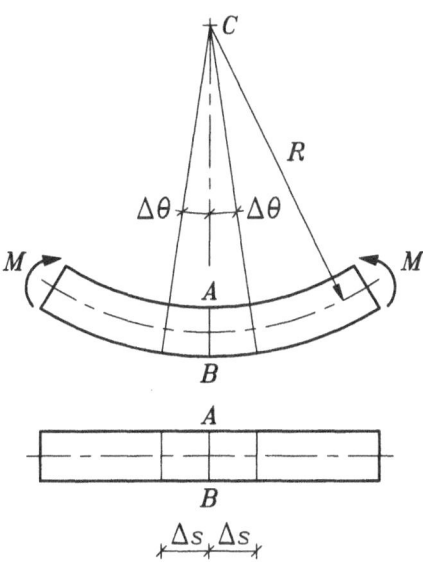

Figure 2.2: Plane bending of homogeneous beam.

deformed state. This can be seen e.g. by imagining the figure viewed from the back side of the paper.

The deformed shape of the beam is determined in the following way. The figure shows two cross-sections at a distance Δs to the left and to the right of AB, respectively. By symmetry the extension of these cross-sections must intersect at a common point $\dot{C}$. The part of the beam closest to C will be shortened and the part on the far side elongated. Thus, there must be a so-called *neutral plane* which retains its initial length along the beam. It is convenient to place the *beam axis* in the plane of symmetry and in the neutral plane. This axis, called the *neutral axis*, is indicated in the figure.

All cross-sections have the same internal forces and thereby the same state of deformation. Therefore the distance from the beam axis to the point C is the same for all sections. Thus, the initially straight beam axis is bent into a circle with center C. The deformed shape is characterized completely by the radius R of this circle. This is essentially the explanation of beam bending given by ROBERT HOOKE in 1678 as shown in Fig. 2.3.

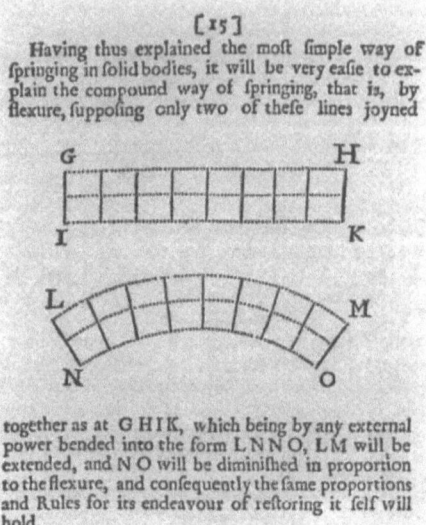

[15]

Having thus explained the moſt ſimple way of ſpringing in ſolid bodies, it will be very eaſie to explain the compound way of ſpringing, that is, by flexure, ſuppoſing only two of theſe lines joyned

together as at G H I K, which being by any external power bended into the form L N N O, L M will be extended, and N O will be diminiſhed in proportion to the flexure, and conſequently the ſame proportions and Rules for its endeavour of reſtoring it ſelf will hold.

Figure 2.3: HOOKE's explanation of beam bending (1678).

In practice it is convenient to describe the deformation by the *curvature* $\kappa = 1/R$, and in order to obtain a quantitative theory the relation between the bending moment M and the curvature κ must be established. Figure 2.4 shows a beam section with initial length Δs. The neutral axis retains its length, and in the deformed state

$$\kappa = \frac{1}{R} = \frac{\Delta\theta}{\Delta s} \tag{2.1}$$

where $\Delta\theta$ is the angle between the two cross-sections. Now, consider a fiber located at the distance z below the neutral axis. The initial length of this fiber is Δs. After deformation it is bent into a circle with radius $R + z$, and corresponds to an angle $\Delta\theta$. Thus its length after deformation is

$$\Delta s_* = (R + z)\,\Delta\theta = (R + z)\,\kappa\,\Delta s \tag{2.2}$$

where the angle $\Delta\theta$ was substituted from (2.1). The elongation corresponds to the normal strain

$$\varepsilon = \frac{\Delta s_* - \Delta s}{\Delta s} = \kappa\,z \tag{2.3}$$

Thus, in plane bending the normal strain is proportional to the curvature κ and to the distance z from the neutral axis.

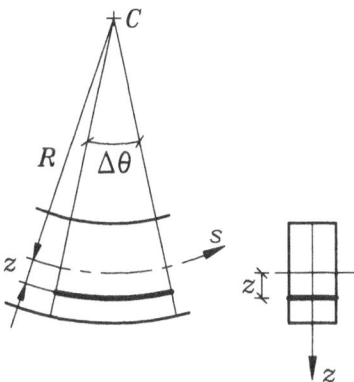

Figure 2.4: Plane bending of rectangular beam section.

If the beam material is linear elastic, the normal stress on a cross-section in bending follows from combining the strain relation (2.3) with Hooke's law (1.14).

$$\sigma = E\varepsilon = E\kappa z \qquad (2.4)$$

This stress distribution leads to a very simple relation between the bending moment M and the curvature κ. The bending moment is determined by integrating the contributions from the stress at each point of the cross-section, multiplied with its distance z from the neutral axis.

$$M = \int_A \sigma z \, \mathrm{d}A = \kappa \int_A E z^2 \, \mathrm{d}A \qquad (2.5)$$

Thus, the moment M is proportional to the curvature κ of the neutral axis. The coefficient of proportionality, representing the bending stiffness of the beam, is the integral of Ez^2 over the cross-section area A. The bending stiffness depends on the stiffness of the material, represented by the elastic modulus E, and the shape and size of the beam cross-section. For a homogeneous beam, i.e. a beam where the elastic modulus E is identical at all points, the bending stiffness relation (2.5) immediately separates into the form

$$M = EI\kappa \qquad (2.6)$$

where

$$I = \int_A z^2 \, \mathrm{d}A \qquad (2.7)$$

is the moment of inertia about the neutral line.

Box 2.1: Beam bending

In a linear elastic beam in plane bending the curvature κ is proportional to the bending moment M.

$$M = \kappa \int_A E\, z^2 \,\mathrm{d}A$$

For a beam with homogeneous material properties this is

$$M = EI\kappa \quad , \quad I = \int_A z^2 \,\mathrm{d}A$$

where I is the bending moment of inertia about the neutral line.

It is clear from the definition (2.7) of the bending moment of inertia of a beam, that it is equal to the area of the cross-section times the square of some 'effective height' of the section. Detailed cross-section analysis is outside the scope of this book, see e.g. Gere & Timoshenko (1997), but typical values are given in the following example.

Example 2.1

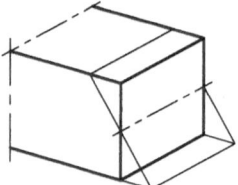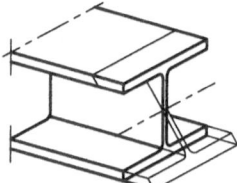

Figure 2.5: Plane bending of rectangular beam section.

Figure 2.5a shows the distribution of the coordinate z over the cross-section of a rectangular beam of height h and width b. The neutral axis is along the axis of symmetry, and the bending moment of inertia of the rectangular cross-section then is

$$I = \int_A z^2 \,\mathrm{d}A = b\int_{-h/2}^{h/2} z^2 \,\mathrm{d}z = \tfrac{1}{12}h^3 b = \tfrac{1}{12}h^2 A$$

Box 2.2: Curvature

The curvature κ of a circle with radius R is defined as

$$\kappa = \frac{1}{R}$$

For a plane curve the curvature $\kappa(s)$ is defined as a function of the arc length s by introducing the angle $\theta(s)$ between the tangent and a fixed direction. The curvature is the rate of change of the angle with arc length,

$$\kappa(s) = \frac{1}{R(s)} = \lim_{\Delta s \to 0} \frac{\Delta\theta}{\Delta s} = \frac{d\theta}{ds}$$

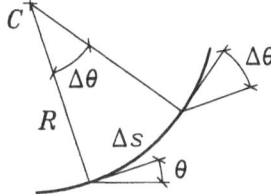

The rectangular beam has the largest bending stiffness, when oriented such that $h \geq b$, as normally seen in structures.

In the I-beam the central part - called the web - is usually thin and contributes only little to the area and bending moment of inertia. The neutral axis is the axis of symmetry, and if the area is assumed to be approximately distributed on the two flanges of thickness t, the bending moment of inertia is

$$I = \int_A z^2 \, dA = 2\left(\tfrac{1}{2}h\right)^2 tb = \tfrac{1}{2}h^2 tb = \tfrac{1}{4}h^2 A$$

It is seen that in the I-beam a given area provides three times greater bending stiffness than in a rectangular section of the same height. The stress distribution is also more favorable in the I-beam, and for materials where the slender web does not pose problems, the I-beam is a common choice for structural use.

For beam bending under constant moment a linear relation has been established between the moment and the curvature κ, Box 2.1. The definition of curvature is easily extended to regular curves as shown in Box 2.2, and a general theory of beam bending can then be established by assuming that the linear relation between moment and curvature, derived for constant moment and curvature, can also be used for non-constant moment and curvature distribution along the beam. This type of beam theory is usually called Bernoulli beam theory. A possible limitation of this theory is the neglect of any effect of shear forces in the beam. However, as demonstrated in Section 2.7 the effects of shear deformation are usually small in slender beams, and Bernoulli beam theory often provides fully satisfactory results.

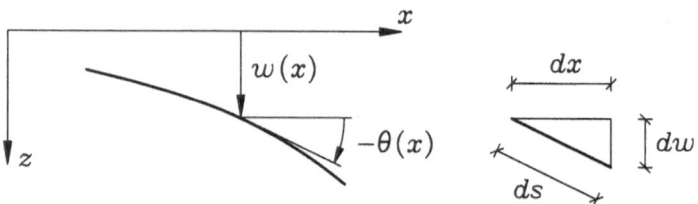

Figure 2.6: Deflected beam axis.

In most cases in practice the curvature - or rather the change in curvature caused by the bending moment - is very small. It is therefore of interest to replace the exact curvature in the beam bending theory with an approximation that leads to a simpler analytical formulation and simpler solutions. The present discussion is limited to straight beams, and the x-axis is used as beam axis in the undeformed state. After deformation the beam axis is described by the curve $(x, w(x))$ in the xz-coordinate system shown in Fig. 2.6. The rotation of the beam axis is described by the angle $\theta(x)$, with counterclockwise rotation positive. The length along the beam axis is denoted s, and it is seen from the small triangle in the figure that

$$\sin \theta = -\frac{dw}{ds} \tag{2.8}$$

When this relation is differentiated with respect to s, it determines the curvature,

$$\kappa = \frac{d\theta}{ds} = -\frac{1}{\cos\theta}\frac{d^2w}{ds^2} = -\frac{\dfrac{d^2w}{ds^2}}{\sqrt{1 - \left(\dfrac{dw}{ds}\right)^2}} \tag{2.9}$$

If the angle θ is small this expression can be simplified considerably. First it is observed that $dx/ds = \cos\theta \simeq 1$, and thus the coordinate x may be used instead of the arc length s also after deformation. Also the factor $\cos\theta$ may be replaced by one in (2.9). As a result of these approximations, valid for $|\theta| << 1$, the rotation and curvature are represented as

$$\theta \simeq -\frac{dw(x)}{dx} \quad , \quad \kappa \simeq \frac{d\theta(x)}{dx} \simeq -\frac{d^2w(x)}{dx^2} \tag{2.10}$$

The approximation involved in using these expressions is illustrated in the following example.

Example 2.2

A common way to establish a constant moment over a section of a beam is the so-called four point bending shown in Fig. 2.7. The load consists of two concentrated forces P acting on the cantilever parts of the beam. This load produces the constant moment $M \simeq Pa$ over the part of the beam between the supports. In a strict sense the moment is only approximately equal to Pa, because the distance from the support may change slightly due to the deformation.

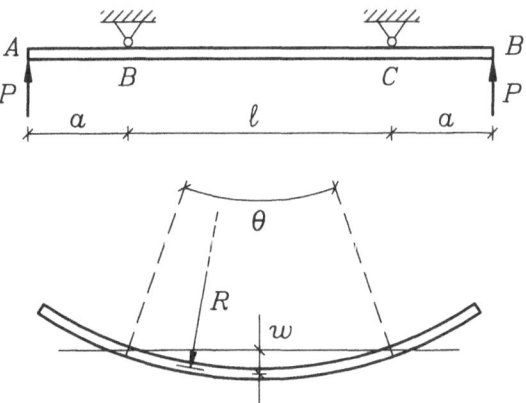

Figure 2.7: Four-point beam bending.

According to the exact beam theory the central part of the beam bends into a circular shape with radius

$$R = \frac{1}{\kappa} = \frac{EI}{M}$$

The angle θ between the cross-sections at the support follows from trigonometry as

$$\sin(\tfrac{1}{2}\theta) = \frac{\ell}{2R} = \tfrac{1}{2}\kappa\ell$$

The Taylor expansion

$$\tfrac{1}{2}\theta = \arcsin(\tfrac{1}{2}\kappa\ell) = (\tfrac{1}{2}\kappa\ell) + \tfrac{1}{6}(\tfrac{1}{2}\kappa\ell)^3 + \cdots$$

shows that for small curvature $\theta \simeq \kappa\ell$.

The displacement w in the middle is

$$w = R(1 - \cos(\tfrac{1}{2}\theta)) = \tfrac{1}{2}R(\tfrac{1}{2}\theta)^2\left(1 - \tfrac{1}{12}(\tfrac{1}{2}\theta)^2 + \cdots\right)$$

where the Taylor expansion of the cos-function has been introduced. The leading term is

$$w = \tfrac{1}{8}R\theta^2 \simeq \tfrac{1}{8}\kappa\ell^2 = \frac{\ell^2\,M}{8EI}$$

This is the result that would be obtained from the approximate theory based on the linearized rotation and curvature relations (2.10), see Exercise 2.2.

The relative error is found by introducing the Taylor expansion of θ into the expansion of w. This leads to the leading relative error term $\tfrac{1}{4}(\tfrac{1}{2}\kappa\ell)^2$. The top of the beam is in compression and the bottom in tension. The curvature can be expressed in terms of the beam height and the maximum strain by use of the relation (2.3). The maximum strain ε_{max} occurs at the distance $\tfrac{1}{2}h$ from the beam axis, and thus $\kappa = 2\varepsilon_{max}/h$. The leading relative error term can then be expressed as

$$\frac{w - w_{lin}}{w_{lin}} \simeq \frac{1}{4}\left(\varepsilon_{max}\frac{\ell}{h}\right)^2$$

Thus, for a limited maximal strain in the material - e.g. $\varepsilon_{max} \simeq 0.002$ - the error in the linearized theory will only be important for very slender beams.

––––––––––

The remainder of this chapter will be concerned with beam theories based on the linearized rotation and curvature expressions (2.10). Most of the theory of columns presented in Chapter 3 will also make use of these relations. However, Section 3.5 contains a modern version of the full nonlinear column problem, originally treated by LEONHARD EULER in 1644.

2.2 Differential equations of beam bending

In most problems of beam bending the displacements are small relative to the dimensions of the beam, and it is therefore acceptable to consider equilibrium with respect to the undeformed geometry of the beam. Figure 2.8 shows a beam with distributed transverse load of intensity $p(x)$. A slice of thickness dx is shown in detail with the corresponding load and internal forces. On the left side the internal forces are Q, M and on the right they may have changed by the increments dQ, dM, giving the values $Q + dQ$ and $M + dM$. In order for the slice to be in equilibrium the vertical projection of all forces must vanish,

$$(Q + dQ) - Q + p\,dx = 0 \tag{2.11}$$

After cancellation of Q and division by dx the differential equation

$$\frac{dQ}{dx} = -p \tag{2.12}$$

is obtained. The sum of moments must also vanish, whereby

$$(M + dM) - M - Q\,dx = 0 \tag{2.13}$$

where higher order terms like $dQdx$ are neglected. Cancellation of M leads to the differential equation

$$\frac{dM}{dx} = Q \tag{2.14}$$

These two first order differential equations must be satisfied irrespective of the material properties of the beam. The shear force can be eliminated, resulting in a second order differential equation for the moment $M(x)$.

$$\frac{d^2M}{dx^2} + p = 0 \tag{2.15}$$

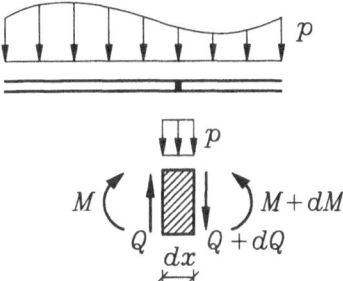

Figure 2.8: Equilibrium of thin beam section.

In some particular cases the boundary conditions can be expressed in terms of the moment M or its derivative $Q = dM/dx$. In these cases the moment distribution $M(x)$ can be determined without considering the deformation properties of the beam. Such a case is called *statically determinate*. The examples treated so far have all been statically determinate. The next two sections deal with internal forces and displacements in statically determinate beams.

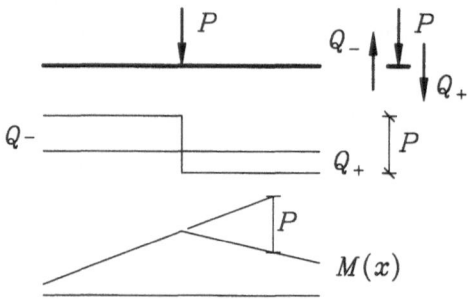

Figure 2.9: Shear force and moment curves at concentrated transverse force.

A concentrated transverse force P leads to a characteristic behaviour of the shear force and moment distributions, as shown in Fig. 2.9. Equilibrium of a small section around the external force gives a discontinuity of the shear force

$$Q_+ - Q_- = -P \qquad (2.16)$$

The moment distribution is continuous at the load, but by (2.14) the derivative of the moment is defined by shear force, and thus the slope of the moment curve exhibits a discontinuity

$$\frac{dM_+}{dx} - \frac{dM_-}{dx} = -P \qquad (2.17)$$

as shown in the figure. Note, that a similar behaviour will be induced by a transverse reaction force.

In beam theories based on small deformations the linearized kinematic relations (2.10) for rotation and curvature are simply used as definitions of θ and κ,

$$\theta(x) = -\frac{dw(x)}{dx} \quad , \quad \kappa(x) = \frac{d\theta(x)}{dx} = -\frac{d^2w(x)}{dx^2} \qquad (2.18)$$

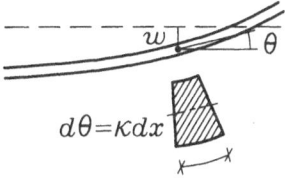

Figure 2.10: Kinematic variables θ and κ.

When this approximate definition of curvature is introduced into the elastic beam bending relation (2.6), the moment is expressed in terms of the second derivative of the displacement.

$$M = EI\kappa = -EI\frac{\mathrm{d}^2 w}{\mathrm{d}x^2} \qquad (2.19)$$

This relation together with the static equilibrium equations form the basis of the so-called Bernoulli beam theory, summarized in Box 2.3. It is seen in Box 2.3 that there is a remarkable similarity between the static equations relating M, Q, p and the kinematic relations between w, θ, κ. In fact, if the variable $-\theta$ were used, the static and kinematic relations would be identical. This complete similarity has been used to devise direct methods similar to those of statics for the solution of beam displacement problems, see e.g. Kassimali (1999). This static analogue of the kinematic equations is called the conjugate beam method.

Statically determinate problems may be solved by first finding the moment distribution $M(x)$ from (2.15), and then solving the curvature equation

$$\frac{\mathrm{d}^2 w}{\mathrm{d}x^2} + \frac{M(x)}{EI} = 0 \qquad (2.20)$$

with appropriate kinematic boundary conditions in terms of w and $\theta = -\mathrm{d}w/\mathrm{d}x$.

If the problem is *statically indeterminate*, it can not be split into a static part, that can be solved independently and then used as input to the kinematic problem. In principle, statically indeterminate problems must be solved by combining the moment equilibrium equation (2.15) and the displacement curvature equation (2.20) into a single 4'th order differential equation,

$$\frac{\mathrm{d}^2}{\mathrm{d}x^2}\left(EI\frac{\mathrm{d}^2 w}{\mathrm{d}x^2}\right) - p(x) = 0 \qquad (2.21)$$

Box 2.3: Bernoulli beam theory

The small displacement elastic Bernoulli beam theory is based on statics (equilibrium),

$$\frac{\mathrm{d}Q}{\mathrm{d}x} = -p \quad , \quad \frac{\mathrm{d}M}{\mathrm{d}x} = Q$$

kinematics,

$$\frac{\mathrm{d}\theta}{\mathrm{d}x} = \kappa \quad , \quad \frac{\mathrm{d}w}{\mathrm{d}x} = -\theta$$

and an elastic relation between moment and curvature,

$$M = EI\,\kappa = -EI\frac{\mathrm{d}^2 w}{\mathrm{d}x^2}$$

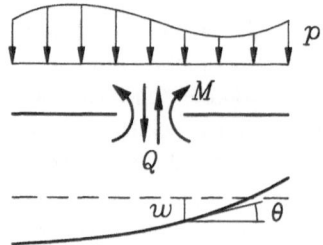

---

and then solving this equation with appropriate boundary conditions formulated in terms of w, $\mathrm{d}w/\mathrm{d}x$, $M = -EI\mathrm{d}^2 w/\mathrm{d}x^2$, and $Q = -\mathrm{d}(EI\mathrm{d}^2 w/\mathrm{d}x^2)/\mathrm{d}x$. However, in practice the solution to linear statically indeterminate problems is more often obtained by using superposition of particular solutions to satisfy the boundary conditions by combination. Some of these techniques are described in Sections 2.5 and 2.6. The differential equations of linear Bernoulli beam theory are summarized in Box 2.4.

The particular form of the 4'th order beam equation, valid also for variable bending stiffness $EI(x)$, should be noted. The structure of this equation is important for the existence of a virtual work principle for deformable beams (Section 2.4), as well as for extensions of beam theory to vibrations (Section 2.9) and stability problems (Chapter 3).

Box 2.4: Differential equations of beam bending

The differential relations of small deformation elastic Bernoulli beam theory from Box 2.3 combine into static and kinematic second order differential equations,

$$\frac{d^2 M}{dx^2} + p(x) = 0 \quad , \quad \frac{d^2 w}{dx^2} + \frac{M(x)}{EI} = 0$$

which can be solved sequentially for statically determinate beams.

For statically indeterminate beams the combined 4'th order beam bending differential equation

$$\frac{d^2}{dx^2}\left(EI\frac{d^2 w}{dx^2}\right) - p(x) = 0$$

must be solved.

Kinematic boundary conditions are in terms of w and dw/dx, and static boundary conditions in terms of

$$M = -EI\frac{d^2 w}{dx^2} \quad , \quad Q = -\frac{d}{dx}\left(EI\frac{d^2 w}{dx^2}\right)$$

2.3 Statically determinate beams

In a statically determinate structure the internal force distribution can be determined from the equilibrium equations alone. For a beam this implies that the moment distribution $M(x)$ can be determined without reference to the displacements. When the moment distribution has been determined, the displacements can be found by integration of the curvature relation (2.20).

This section illustrates the direct integration of the curvature relation. The immediate result is the displacement distribution $w(x)$, and the rotation distribution $\theta(x)$ then follows from differentiation. The rotations θ_A and θ_B at the end points are of particular interest, as they are used in the solution of statically indeterminate beam problems in Section 2.5. The solutions obtained in this section can also be obtained from the principle of virtual work, by the method described in the next section.

Example 2.3

Consider the simply supported beam shown in Fig. 2.11. The length of the beam is ℓ and the bending stiffness EI. The beam is loaded by a concentrated moment M_B at the support B.

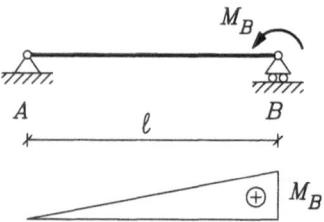

Figure 2.11: Simply supported beam with concentrated end moment.

The beam is statically determinate, and the moment distribution can therefore be found directly from static equilibrium.

$$M(x) = \frac{x}{\ell} M_B$$

The displacement of the beam is found from the bending differential equation (2.20),

$$\frac{\mathrm{d}^2 w}{\mathrm{d}x^2} = -\frac{M(x)}{EI} = -\frac{x}{\ell}\frac{M_B}{EI}$$

and the boundary conditions $w(0) = w(\ell) = 0$, corresponding to simple supports.

The general solution to the differential equation is

$$w = -\frac{1}{6}\frac{x^3}{\ell}\frac{M_B}{EI} + C_0 + C_1 x$$

The two arbitrary constants C_0 and C_1 are determined from the boundary conditions. The condition $w(0)=0$ directly gives $C_0=0$, and $w(\ell)=0$ then gives $C_1 = \dfrac{\ell M_B}{6EI}$. With these constants the displacement of the beam is

$$w(x) = \frac{\ell^2 M_B}{6EI}\frac{x}{\ell}\left[1 - \left(\frac{x}{\ell}\right)^2\right]$$

The first factor gives the 'scale of the displacement', while the following non-dimensional factors give the distribution along the beam.

The rotation of the beam axis follows from differentiation,

$$\theta(x) \;=\; -\frac{dw}{dx} \;=\; -\frac{\ell M_B}{6EI}\Big[1 - 3\Big(\frac{x}{\ell}\Big)^2\Big]$$

This gives the rotation at the end points

$$\theta_A \;=\; -\frac{\ell M_B}{6EI} \quad , \qquad \theta_B \;=\; \frac{\ell M_B}{3EI}$$

The latter of these relations is similar to a rotation spring stiffness k_B for the moment M_B at B,

$$M_B \;=\; k_B\,\theta_B \quad , \qquad k_B \;=\; \frac{3EI}{\ell}$$

The rotation spring stiffness parameter k_B has the dimension of a moment. It is proportional to the bending stiffness of the beam EI and inversely proportional to the beam length ℓ.

A scaled form of this solution with imposed rotation $\theta_B = 1$ is included as case A.5.1 in Appendix A.

Example 2.4

Figure 2.12 shows a simply supported beam of length ℓ and constant bending stiffness EI, loaded by a uniform distributed load of intensity p. The moment is determined from statics, see e.g. Example 1.3,

$$M(x) \;=\; \tfrac{1}{2}\,p\,x(\ell - x)$$

This incorporates the boundary conditions $M(0) = M(\ell) = 0$, corresponding to the simple supports.

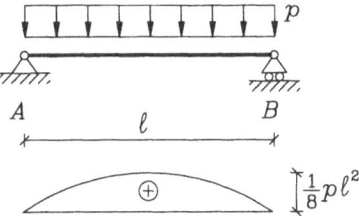

Figure 2.12: Simply supported beam with uniformly distributed load.

The differential equation (2.20) for the beam displacement $w(x)$ then is

$$\frac{\mathrm{d}^2 w}{\mathrm{d}x^2} = -\frac{M(x)}{EI} = -\frac{p}{2EI} x(\ell - x)$$

with the boundary conditions $w(0) = w(\ell) = 0$. Integration gives

$$\frac{\mathrm{d}w}{\mathrm{d}x} = -\frac{p}{2EI}\left(\tfrac{1}{2}\ell x^2 - \tfrac{1}{3}x^3 + C_1\right)$$

and by integrating once more

$$w = -\frac{p}{2EI}\left(\tfrac{1}{6}\ell x^3 - \tfrac{1}{12}x^4 + C_1 x + C_2\right)$$

The arbitrary constant $C_2 = 0$ is determined directly by the boundary condition $w(0)=0$. The remaining constant C_1 then follows from $w(\ell)= 0$,

$$\tfrac{1}{6}\ell^4 - \tfrac{1}{12}\ell^4 + C_1 \ell = 0$$

whereby $C_1 = -\tfrac{1}{12}\ell^3$. The displacement has then been determined as

$$w(x) = \frac{p\,\ell^4}{24EI}\frac{x}{\ell}\left[1 - 2\left(\frac{x}{\ell}\right)^2 + \left(\frac{x}{\ell}\right)^3\right]$$

The rotation angle is found as

$$\theta(x) = -\frac{\mathrm{d}w}{\mathrm{d}x} = -\frac{p\,\ell^3}{24EI}\left[1 - 6\left(\frac{x}{\ell}\right)^2 + 4\left(\frac{x}{\ell}\right)^3\right]$$

either by substituting C_1 into the expression for $\mathrm{d}w/\mathrm{d}x$ or by differentiation of $w(x)$. In particular the rotations at the ends of the beam are

$$\theta_B = -\theta_A = \frac{p\,\ell^3}{24EI}$$

This solution is summarized as case A.1.2 in Appendix A.

In the examples 2.3 and 2.4 the integration of the curvature relation to obtain the displacement distribution $w(x)$ was quite straightforward, because the load was described by a smooth simple function over the full length of the beam, and the bending stiffness EI was constant along the beam. The following example treats the slightly more complicated case of a concentrated load. Problems involving non-constant bending stiffness are often treated more effectively by use of the principle of virtual work, described in Section 2.4.

Example 2.5

Figure 2.13 shows a simply supported beam of length ℓ and constant bending stiffness EI, loaded by a concentrated transverse force P, located at the distance a from the left end point A. The notation $a' = \ell - a$ is introduced for the distance from the right end point B. Similarly the displacement is described as a function of the coordinate x or $x' = \ell - x$.

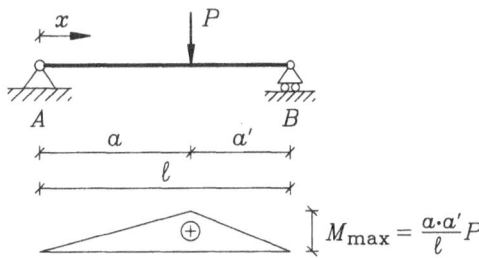

Figure 2.13: Simply supported beam with concentrated force.

The beam is statically determinate with the maximum moment under the force, $M_{max} = Paa'/\ell$. The analytical expressions for the moment distribution are

$$M(x) \;=\; \begin{cases} x\,\dfrac{a'}{\ell}\,P & , \quad x \le a \\[2mm] x'\,\dfrac{a}{\ell}\,P & , \quad x \ge a \end{cases}$$

The curvature relation (2.20) can be expressed in either of the variables x or x',

$$\frac{\mathrm{d}^2 w}{\mathrm{d}x^2} = -\frac{M(x)}{EI} \qquad \text{or} \qquad \frac{\mathrm{d}^2 w}{\mathrm{d}x'^2} = -\frac{M(x')}{EI}$$

Integration of these equations for the left and right part of the beam then gives

$$w(x) \;=\; \begin{cases} -\dfrac{x^3 a'}{6\ell}\dfrac{P}{EI} + C_0 + C_1\, x & , \quad x \le a \\[3mm] -\dfrac{x'^3 a}{6\ell}\dfrac{P}{EI} + D_0 + D_1\, x' & , \quad x \ge a \end{cases}$$

The boundary conditions $w = 0$ for $x = 0$ and $x = \ell$ determine the arbitrary constants $C_0 = D_0 = 0$. The remaining constants C_1 and D_1

are to be determined from the requirement that $w(x)$ and $dw(x)/dx$ are continuous at $x = a$ (and $x' = a'$). Continuity of $w(x)$ gives by direct substitution

$$-\frac{a^3 a'}{6\ell}\frac{P}{EI} + C_1 a = -\frac{a'^3 a}{6\ell}\frac{P}{EI} + D_1 a'$$

while continuity of $dw/dx = -dw/dx'$ follows from differentiation of the left and right part solutions,

$$-3\frac{a^2 a'}{6\ell}\frac{P}{EI} + C_1 = 3\frac{a'^2 a}{6\ell}\frac{P}{EI} - D_1$$

Elimination of D_1 by multiplication of the last equation with a' gives

$$-(a^2 + 3aa')\frac{aa'}{6\ell}\frac{P}{EI} + (a + a')C_1 = 2a'^2\frac{a'a}{6\ell}\frac{P}{EI}$$

Now, $a + a' = \ell$, and thus

$$C_1 = (a^2 + 3aa' + 2a'^2)\frac{a'a}{6\ell^2}\frac{P}{EI}$$

This expression simplifies considerably by substitution of $a = \ell - a'$ in the first parenthesis,

$$C_1 = (\ell + a')\frac{a'a}{6\ell}\frac{P}{EI} = (\ell^2 - a'^2)\frac{a'}{6\ell}\frac{P}{EI} = \left[1 - \left(\frac{a'}{\ell}\right)^2\right]\frac{P\ell a'}{6EI}$$

Due to symmetry of the formulation the constant D_1 is found by replacement of a' by a.

$$D_1 = \left[1 - \left(\frac{a}{\ell}\right)^2\right]\frac{P\ell a}{6EI}$$

The final solution is obtained by substitution of the constants C_1 and D_1 into the left and right solution, respectively.

$$w(x) = \begin{cases} \dfrac{P\ell a' x}{6EI}\left[1 - \left(\dfrac{a'}{\ell}\right)^2 - \left(\dfrac{x}{\ell}\right)^2\right] & , \ x \le a \\[4mm] \dfrac{P\ell a x'}{6EI}\left[1 - \left(\dfrac{a}{\ell}\right)^2 - \left(\dfrac{x'}{\ell}\right)^2\right] & , \ x \ge a \end{cases}$$

This solution is included as case A.1.1 in Appendix A.

2.4 Virtual work and beam displacements

It was demonstrated in Section 1.5 that the equilibrium conditions of a force system, e.g. the forces acting on a body, are equivalent to the statement, that the virtual work of all forces and moments under an arbitrary virtual rigid body motion must vanish. For a rigid body the concept of virtual work is quite simple. In fact, it corresponds to replacing a force projection equation with a virtual translation in this direction, and replacing a moment equation about an axis with a virtual rotation about this axis.

It was seen in Example 1.5 that for beams connected by hinges, the virtual work principle remains valid, if each beam was given its own virtual rigid body displacement, without breaking the connections at the hinges. The reason is, that the work of the internal force pairs at the hinges cancel, because their virtual displacements are identical, and the internal moment vanishes at a hinge. Thus, there is no contribution to the virtual work from the internal forces at the hinge.

The principle of virtual work can be extended to flexible bodies, using more general virtual displacement fields corresponding to deformation of the body. To be specific, consider a straight beam under transverse load $p(x)$ with moment distribution $M(x)$. The moment distribution must satisfy the equilibrium equation (2.15),

$$\frac{d^2 M}{dx^2} + p(x) = 0 \qquad (2.22)$$

This is an equation for the transverse load intensity, and the work performed through a virtual transverse displacement field $\delta w(x)$ is

$$\int_0^\ell \delta w(x) \left(\frac{d^2 M}{dx^2} + p(x) \right) dx - 0 \qquad (2.23)$$

The equilibrium equation (2.22) is valid for any value of x, and the virtual work integral must therefore vanish for any choice of the virtual displacement field $\delta w(x)$. The first term in the parenthesis is a second order derivative. If the virtual displacement field $\delta w(x)$ and its first derivative $d(\delta w(x))/dx$ are continuous, the double differentiation of $M(x)$ can be removed via two integrations by parts of the first term. The equation then becomes

$$\left[\delta w\, Q - \frac{d(\delta w)}{dx} M \right]_0^\ell + \int_0^\ell \left(\frac{d^2(\delta w)}{dx^2} M + \delta w\, p(x) \right) dx = 0 \qquad (2.24)$$

This equation takes a more familiar form when the rotation and curvature of the virtual displacement field $\delta w(x)$ are introduced according to the linearized definitions (2.18).

$$\delta\theta(x) \; = \; -\frac{d(\delta w(x))}{dx} \quad , \quad \delta\kappa(x) \; = \; \frac{d(\delta\theta(x))}{dx} \; = \; -\frac{d^2(\delta w(x))}{dx^2} \qquad (2.25)$$

When these definitions of the virtual rotation $\delta\theta$ and the virtual curvature $\delta\kappa$ are introduced into (2.24), the following equation is obtained

$$\left[\delta w\, Q \; + \; \delta\theta\, M \right]_0^\ell \; + \; \int_0^\ell \delta w\, p(x)\; dx \; = \; \int_0^\ell \delta\kappa\, M\; dx \qquad (2.26)$$

This is the virtual work equation of a beam with distributed transverse load $p(x)$, and a virtual displacement field with continuous $\delta w(x)$ and $\delta\theta(x)$. The left side of the equation is the virtual work of the distributed external load $p(x)$ and the forces and moments at the beam ends. These terms are similar to those of the rigid body version of the virtual work presented in Section 1.5. However, in the present case the external virtual work generally does not vanish. If the virtual displacement field $\delta w(x)$ does not correspond to a rigid body motion of the beam, there will be a virtual curvature $\delta\kappa(x)$, and the right side of the equation represents the *internal virtual work* of the moment $M(x)$ through the virtual incremental angle $d(\delta\theta) = \delta\kappa dx$, shown in Fig. 2.10.

The virtual work equation (2.26) for a beam is easily generalized to include concentrated loads P_i and M_j. At the points, where the concentrated loads act, the internal forces satisfy the discontinuity conditions

$$Q(x_i^+) - Q(x_i^-) \; = \; -P_i \quad , \qquad M(x_j^+) - M(x_j^-) \; = \; -M_j \qquad (2.27)$$

The continuity conditions on the virtual displacement field can also be relaxed by permitting discontinuities

$$\delta w(x_k^+) - \delta w(x_k^-) \; = \; \Delta w_k \quad , \qquad \delta\theta(x_l^+) - \delta\theta(x_l^-) \; = \; \Delta\theta_l \qquad (2.28)$$

The general form of the virtual work equation is now obtained by integration by parts between the points of discontinuity. With the standard sign convention the internal forces at $x = 0$ are the negative of the corresponding loads, $Q(0) = -P_0$ and $M(0) = -M_0$. When the shear force and moment at the ends of the beam are included among the external loads, the virtual work equation (2.26) takes the form

$$\Sigma_i \, \delta w_i \, P_i \; + \; \Sigma_j \, \delta\theta_j \, M_j \; + \; \int_0^\ell \delta w\, p \; dx$$
$$= \; \Sigma_k \, \Delta w_k \, Q(x_k) \; + \; \Sigma_l \, \Delta\theta_l \, M(x_l) \; + \; \int_0^\ell \delta\kappa\, M\; dx \qquad (2.29)$$

Box 2.5: Virtual work equation for Bernoulli beam

A beam is acted on by the distributed transverse load $p(x)$ and concentrated loads P_i, M_j, including reactions. The internal moment $M(x)$ and shear force $Q(x)$ then satisfy the static differential relations

$$\frac{\mathrm{d}Q}{\mathrm{d}x} = -p \quad , \qquad \frac{\mathrm{d}M}{\mathrm{d}x} = Q$$

and discontinuity relations at the concentrated loads

$$Q(x_i^+) - Q(x_i^-) = -P_i \quad , \qquad M(x_j^+) - M(x_j^-) = -M_j$$

A virtual displacement field $\delta w(x)$ defines the virtual curvature and rotation as

$$\frac{\mathrm{d}(\delta\theta)}{\mathrm{d}x} = \delta\kappa \quad , \qquad \frac{\mathrm{d}(\delta w)}{\mathrm{d}x} = -\delta\theta$$

with possible discontinuities

$$\delta w(x_k^+) - \delta w(x_k^-) = \Delta w_k \quad , \qquad \delta\theta(x_l^+) - \delta\theta(x_l^-) = \Delta\theta_l$$

For an arbitrary choice of the virtual displacement field $\delta w(x)$ the external virtual work is equal to the internal virtual work,

$$\sum_i \delta w_i\, P_i \; + \; \sum_j \delta\theta_j\, M_j \; + \; \int_0^\ell \delta w\, p\, \mathrm{d}x$$

$$= \; \sum_k \Delta w_k\, Q(x_k) \; + \; \sum_l \Delta\theta_l\, M(x_l) \; + \; \int_0^\ell \delta\kappa\, M\, \mathrm{d}x$$

The left side of this equation is identified as the *external virtual work*, i.e. the virtual work of all externally applied forces and moments,

$$\delta W_{ex} \; = \; \sum_i \delta w_i\, P_i \; + \; \sum_j \delta\theta_j\, M_j \; + \; \int_0^\ell \delta w\, p(x)\, \mathrm{d}x \qquad (2.30)$$

and the right side is identified as the *internal virtual work*,

$$\delta W_{in} \; = \; \sum_k \Delta w_k\, Q(x_k) \; + \; \sum_l \Delta\theta_l\, M(x_l) \; + \; \int_0^\ell \delta\kappa\, M\, \mathrm{d}x \qquad (2.31)$$

Thus, for a flexible beam the principle of virtual work states that the external virtual work δW_{ex} is equal to the internal virtual work δW_{in} for an arbitrary virtual displacement field $\delta w(x)$ with rotation, curvature and discontinuities as defined by (2.25) and (2.28). The principle of virtual work of Bernoulli beams is summarized in Box 2.5.

The internal virtual work represents the work of the internal forces and moments through the virtual deformation of the beam. In the present beam theory there is only one deformation mechanism, namely beam curvature, and the distributed internal work is the work of the internal moment through the virtual curvature $\delta\kappa$, represented as $\delta\kappa M$ in the integral. A virtual displacement field with one or more discontinuities may also be selected. The discontinuities contribute work through the internal forces at the point of discontinuity. Thus, a discontinuity $\Delta\theta_l$ of the virtual rotation corresponds to a concentrated curvature, contributing the virtual work $\Delta\theta_l M(x_l)$, while a virtual displacement discontinuity Δw_k corresponds to a transverse separation, contributing the virtual work $\Delta w_k Q(x_k)$. Discontinuous virtual displacement fields can be used to calculate the shear force or moment in a beam in much the same way that virtual rigid body displacements were used to calculate reactions in Section 1.5. However, further discussion of this falls outside the main scope of this text, and reference is given to books on statics of structures, e.g. Felton & Nelson (1997) and Kassimali (1999).

The principle of virtual work is often used in a slightly different form, in which the roles of the actual and virtual fields are interchanged. In the principle of virtual work, stated in Box 2.5, the static fields $M(x), Q(x)$ are those actually present in the structure, when loaded by the distributed force $p(x)$ and the concentrated loads P_i, M_j. The virtual fields $\delta\kappa(x), \delta\theta(x)$ and the discontinuities $\Delta w_k, \Delta\theta_l$ are simply derived from a selected virtual displacement field $\delta w(x)$. Thus, the principle of virtual work is a principle involving the *actual static fields*, and *virtual kinematic fields*. However, for elastic structures the kinematic fields of the actual structure can be calculated directly from the static fields, and the virtual displacement field may be selected as the displacement corresponding to some virtual loading of the structure. The implication is, that the roles of static and kinematic fields can be interchanged. The corresponding principle is called the *principle of complementary virtual work*. It can be identified directly from the relations in Box 2.5, but here an independent derivation is given to illustrate the similarity between the static and the kinematic equations of beam theory. The various work principles in structural and solid mechanics and the relations between them have been discussed in detail by Washizu (1974).

The principle of complementary virtual work is constructed from the actual displacements of the beam. In beam theory with linearized kinematics the curvature $\kappa(x)$ is given in terms of the displacement $w(x)$ by the equation

$$\frac{d^2 w}{dx^2} + \kappa(x) = 0 \qquad (2.32)$$

This identity is multiplied with a virtual moment field to form the complementary work integral

$$\int_0^\ell \delta M(x)\left(\frac{d^2 w}{dx^2} + \kappa(x)\right) dx = 0 \qquad (2.33)$$

The first term is integrated by parts two times, whereby

$$\left[-\delta M\,\theta - \frac{d(\delta M)}{dx}\,w \right]_0^\ell + \int_0^\ell \left(\frac{d^2(\delta M)}{dx^2}\,w + \delta M\,\kappa\right) dx = 0 \qquad (2.34)$$

where the rotation is defined by $\theta = -dw/dx$. Now, introduce the static equations (2.12) and (2.14) of the virtual moment distribution,

$$\frac{d(\delta Q)}{dx} = -\delta p \quad , \qquad \frac{d(\delta M)}{dx} = \delta Q \qquad (2.35)$$

whereby the complementary virtual work equation (2.34) takes the form

$$\left[\delta M\,\theta + \delta Q\,w \right]_0^\ell + \int_0^\ell \delta p\,w\,dx = \int_0^\ell \delta M\,\kappa\,dx \qquad (2.36)$$

Note, that this statement corresponds *exactly* to the equation of virtual work (2.26), when the static fields $M(x), Q(x), p(x)$ are interchanged with the kinematic fields $w(x), -\theta(x), \kappa(x)$ and vice versa. In fact, the derivation also follows identical steps. The actual displacement and rotation of the beam $w(x), \theta(x)$ are continuous, and therefore the only discontinuities to be incorporated are concentrated virtual forces δP_i and moments δM_j. After using discontinuity relations similar to (2.27) the final form of the equation of complementary virtual work is

$$\Sigma_i\,\delta P_i\,w_i + \Sigma_j\,\delta M_j\,\theta_j + \int_0^\ell \delta p\,w\,dx = \int_0^\ell \delta M\,\kappa\,dx \qquad (2.37)$$

This complementary virtual work equation is analogous to the virtual work equation (2.29). It is summarized in Box 2.6.

The complementary virtual work equation (2.37) has an important application in the calculation of the displacement or rotation at a specific point of

Box 2.6: Complementary virtual work

Let a beam have the displacement field $w(x)$ and continuous rotation and curvature fields $\theta(x), \kappa(x)$, defined by

$$\frac{d\theta}{dx} = \kappa \quad , \quad \frac{dw}{dx} = -\theta$$

Introduce a virtual moment field $\delta M(x)$ in equilibrium with the virtual shear force $\delta Q(x)$ and virtual load $\delta p(x)$,

$$\frac{d(\delta Q)}{dx} = -\delta p \quad , \quad \frac{d(\delta M)}{dx} = \delta Q$$

and the concentrated virtual forces δP_i and virtual moments δM_j. The actual kinematic fields and the virtual static fields then satisfy the complementary virtual work equation

$$\Sigma_i \, \delta P_i \, w_i + \Sigma_j \, \delta M_j \, \theta_j + \int_0^\ell \delta p \, w \, dx = \int_0^\ell \delta M \, \kappa \, dx$$

If the virtual moment field corresponds to a single virtual force δP_i, the displacement w_i is given by the integral

$$\delta P_i \, w_i = \int_0^\ell \delta M \, \kappa \, dx = \int_0^\ell \frac{\delta M \, M}{EI} \, dx$$

and if the virtual moment field corresponds to a single virtual moment δM_j the rotation θ_j is given by

$$\delta M_j \, \theta_j = \int_0^\ell \delta M \, \kappa \, dx = \int_0^\ell \frac{\delta M \, M}{EI} \, dx$$

a beam. If the virtual moment field $\delta M(x)$ corresponds to a concentrated virtual load, the distributed virtual load $\delta p(x)$ vanishes, and the equation of complementary virtual work becomes an explicit expression for the displacement or rotation in the form of an integral. If the virtual moment field $\delta M(x)$ corresponds to a single virtual force δP_i the result is

$$\delta P_i \, w_i = \int_0^\ell \delta M \, \kappa \, dx = \int_0^\ell \frac{\delta M \, M}{EI} \, dx \qquad (2.38)$$

where the last form follows from the curvature relation (2.6) for an elastic beam. For a unit virtual force $\delta P_i = 1$ the displacement w_i is given directly by the integral. In a similar way a virtual moment field $\delta M(x)$ corresponding to a concentrated virtual moment δM_j determines the rotation θ_j via the integral

$$\delta M_j\, \theta_j \;=\; \int_0^\ell \delta M\, \kappa\, dx \;=\; \int_0^\ell \frac{\delta M\, M}{EI}\, dx \tag{2.39}$$

Note, that the relations (2.38) and (2.39) are also valid for variable or discontinuous bending stiffness, $EI(x)$.

In many cases the rotation at the end of a simply supported beam is of particular interest, because it determines how much support can be mobilized by connecting the beam rigidly to e.g. a side span. Often the full deformation of the beam is not required, and it is then particularly relevant to determine the rotations by use of the principle of complementary virtual work, by which this limited information can be obtained directly, without solving for the full deformation. The procedure is illustrated in the following example.

Example 2.6

Figure 2.14 shows a simply supported beam with uniform transverse load of intensity p. First the internal moment $M(x)$ is determined. In the present case the internal moment is determined directly from statics, giving the parabolic distribution

$$M(x) \;=\; \tfrac{1}{2} p\, x(\ell - x)$$

with maximum value $M(\tfrac{1}{2}\ell) = \tfrac{1}{8}p\ell^2$.

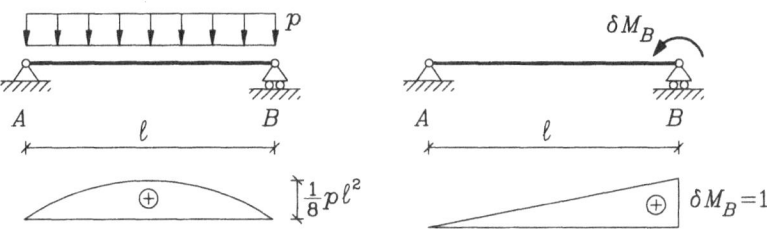

Figure 2.14: Simply supported beam with: a) uniformly distributed load, b) concentrated virtual unit moment $\delta M_B = 1$.

In order to determine the rotation θ_B at B a virtual unit moment $\delta M_B = 1$ is applied at B. The corresponding internal virtual moment distribution is

$$\delta M(x) \;=\; \frac{x}{\ell}\,\delta M_B \;=\; \frac{x}{\ell}$$

i.e. the linear variation shown in the figure.

Now, the principle of complementary virtual work is used for the virtual moment field and the actual displacement field.

$$\delta M_B\,\theta_B \;=\; \int_0^\ell \delta M(x)\,\kappa(x)\;\mathrm{d}x \;=\; \int_0^\ell \delta M(x)\,\frac{M(x)}{EI}\;\mathrm{d}x$$

where the relation $\kappa(x) = M(x)/EI$ is used to express the actual curvature distribution in terms of the moment distribution $M(x)$. The virtual moment δM_B is normalized to unity, and thus the rotation is given directly by the integral,

$$\theta_B \;=\; \frac{1}{EI}\int_0^\ell \delta M(x)\,M(x)\;\mathrm{d}x \;=\; \frac{1}{EI}\,\frac{2\ell}{3}\,\frac{1}{2}\,\frac{p\,\ell^2}{8}$$

where the evaluation is carried out directly as $\frac{1}{2}$ times the area of a parabola, e.g. using the integration formula in Appendix B. The final result is

$$\theta_B \;=\; -\theta_A \;=\; \frac{1}{24}\,\frac{p\,\ell^3}{EI}$$

where θ_A follows from symmetry. This result was obtained in Example 2.4 by differentiation of the complete solution.

The principle of complementary virtual work may also be used to find the displacement of the beam at a particular point by applying a normalized virtual force $\delta P = 1$ at this point, as shown in the following example.

Example 2.7

The displacement at the center of the uniformly loaded beam of the previous example is determined by the principle of complementary virtual work by considering the virtual moment field $\delta M(x)$ corresponding to a concentrated transverse virtual unit force $\delta P = 1$, acting at the center of the beam as shown in Fig. 2.15.

Now, the principle of complementary virtual work is used for the virtual moment field and the actual displacement field.

$$\delta P\,w(\tfrac{1}{2}\ell) \;=\; \int_0^\ell \delta M(x)\,\kappa(x)\;\mathrm{d}x \;=\; \int_0^\ell \delta M(x)\,\frac{M(x)}{EI}\;\mathrm{d}x$$

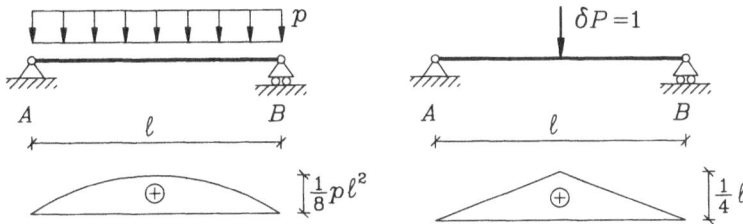

Figure 2.15: Simply supported beam with: a) uniformly distributed
load, b) concentrated virtual unit force $\delta P = 1$.

where the moment fields $M(x)$ and $\delta M(x)$ are shown in the figure. The
magnitude of the virtual force is unity, and thus the deflection at the
center is determined directly by the integral

$$w(\tfrac{1}{2}\ell) \;=\; \frac{1}{EI}\int_0^\ell \delta M(x)\,M(x)\,\mathrm{d}x \;=\; \frac{2}{EI}\,\frac{\ell}{2}\,\frac{5}{12}\,\frac{p\ell^2}{8}\,\frac{\ell}{4} \;=\; \frac{5}{384}\,\frac{p\,\ell^4}{EI}$$

The integral is evaluated using the integration formula in Appendix B.
The same result follows from the complete displacement function $w(x)$
determined in Example 2.4.

In the previous two examples the principle of complementary virtual work
was used to find a rotation or a displacement at a particular point of the
beam. In fact, the principle of complementary virtual work can also be used
to determine the complete displacement field $w(x)$ by applying a unit virtual
force $\delta P = 1$ at the generic point x, as shown in the following example.

Example 2.8

Figure 2.16 shows a cantilever beam of length ℓ and bending stiffness
EI, loaded by a concentrated force P at the distance a from the support.
The moment field $M(x)$ is shown in Fig. 2.16a.

The displacement $w(x)$ at the distance x from the support is determined by placing a virtual unit force $\delta P = 1$ at this point as shown in
Fig. 2.16b. By the principle of complementary virtual work

$$\delta P\,w(x) \;=\; \int_0^\ell \delta M(s)\,\kappa(s)\,\mathrm{d}s \;=\; \int_0^\ell \delta M(s)\,\frac{M(s)}{EI}\,\mathrm{d}s$$

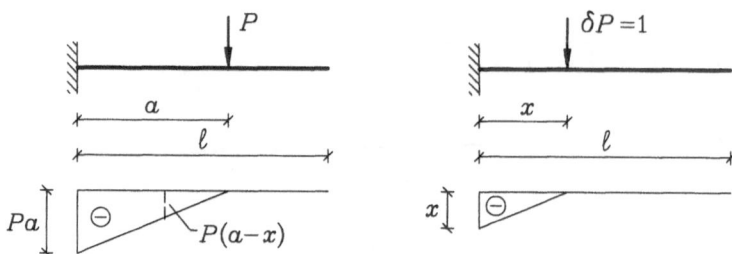

Figure 2.16: Cantilever beam with: a) concentrated actual force P,
b) concentrated virtual unit force $\delta P = 1$.

and thus, for constant bending stiffness and unit magnitude of the virtual
force, the displacement $w(x)$ is given explicitly by the integral

$$w(x) \; = \; \frac{1}{EI} \int_0^\ell \delta M(s)\, M(s) \; \mathrm{d}s$$

It is seen from the moment curves in the figure, that the integrand only
takes non-zero values in the interval from 0 to $\min(a, x)$.

The integral is first evaluated for the case $x \leq a$. In this case the
displacement is evaluated from the integral

$$w(x) \; = \; \frac{1}{EI} \int_0^x \delta M(s)\, M(s) \; \mathrm{d}s \; = \; \frac{1}{EI}\, \frac{x}{6} \Big(2(Pa)x \, + \, P(a - x)\, x \Big)$$

by using the integration formula in Appendix B for the product of a
trapezoidal and a triangular distribution.

$$w(x) \; = \; \frac{Pax^2}{6EI} \Big(3 - \frac{x}{a} \Big) \qquad , \qquad \text{for } x \leq a$$

For $x \geq a$ the roles of a and x are interchanged, and the result can be
written down directly,

$$w(x) \; = \; \frac{Pa^3}{6EI} \Big(3\frac{x}{a} - 1 \Big) \qquad , \qquad \text{for } x \geq a$$

Note, that the deflection is linear for $x \geq a$, corresponding to the beam
remaining straight outside the applied force. The result is summarized
as case A.2.1 in Appendix A.

The principles of virtual and complementary virtual work are not limited to constant bending stiffness. Thus, EI may vary along the beam and even exhibit discontinuities. For statically determinate beams the moment distribution $M(x)$ is independent of the bending stiffness of the beam, and the complementary virtual energy integrals (2.38) and (2.39) for the displacement and rotation therefore only change by the change in the bending flexibility factor $(EI)^{-1}$. This is illustrated in the following example, where the effect of reducing the bending stiffness for the outer half of a cantilever beam is calculated.

Example 2.9

Figure 2.17 shows a cantilever beam of length ℓ, loaded by a transverse force P at the tip. If the bending stiffness is constant of magnitude EI along the full beam length, the displacement at the tip was found in Example 2.8 to be $w(\ell) = P\ell^3/3EI$. Now, the bending stiffness in the outer half of the beam is reduced to $\frac{1}{2}EI$, and the question is how much this increases the tip displacement.

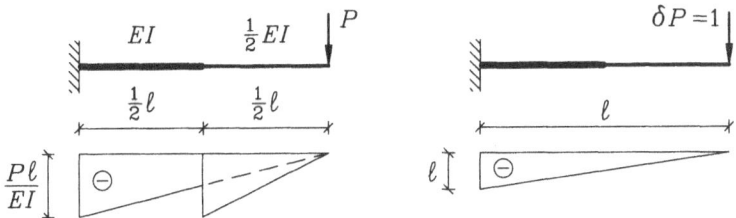

Figure 2.17: Cantilever beam with bending stiffness EI and $\frac{1}{2}EI$.
a) actual curvature, b) virtual moment field.

The left part of the figure shows the actual curvature, $\kappa(x) = M(x)/EI$ for $x < \frac{1}{2}\ell$ and $\kappa(x) = 2M(x)/EI$ for $x > \frac{1}{2}\ell$. Thus, the curvature is doubled in the outer half of the beam. The displacement at $x = \ell$ follows from the complementary virtual work equation (2.39) in the form

$$\delta P\, w(\ell) = \int_0^\ell \delta M\, \kappa\, dx = \frac{1}{EI}\int_0^{\ell/2} \delta M\, M\, dx + \frac{2}{EI}\int_{\ell/2}^\ell \delta M\, M\, dx$$

In practice it is convenient to evaluate the integral as the product of two triangular distributions over the full length of the beam, as indicated by the dotted line, plus an extra product of triangular distributions over the outer half of the beam. Using the formula for the product of two triangular distributions, given in Appendix B, the result is

$$w(\ell) = \int_0^\ell \delta M \, \kappa \, dx = \frac{1}{3}\ell\frac{P\ell}{EI}\ell + \frac{1}{3}\tfrac{1}{2}\ell\frac{P\ell}{2EI}\tfrac{1}{2}\ell = (1+\tfrac{1}{8})\frac{P\ell^3}{3EI} = \frac{3P\ell^3}{8EI}$$

Thus, the relative increase of the displacement is $\frac{1}{8}$. It is seen that the principle of the calculation remains unchanged, only the bookkeeping increases.

Finally, it should be noted that the principles of virtual work and complementary virtual work are not restricted to individual straight beams. The beams may be joined, e.g. rigidly or by hinges, to form structures. With suitable definition of any concentrated virtual work at the joints both principles remain valid. In particular this implies that the principle of complementary virtual work can also be used to determine displacements and rotations in frame structures. This is illustrated in the following very simple example. A more detailed discussion can be found in texts on static analysis of structures, e.g. Felton & Nelson (1997) and Kassimali (1999).

Example 2.10

Figure 2.18 shows an angle beam ABC consisting of two straight parts AB and BC of length a with constant bending stiffness EI. The beam is loaded by a horizontal force P at C. This example illustrates the calculation of the horizontal displacement w_C and the rotation θ_B.

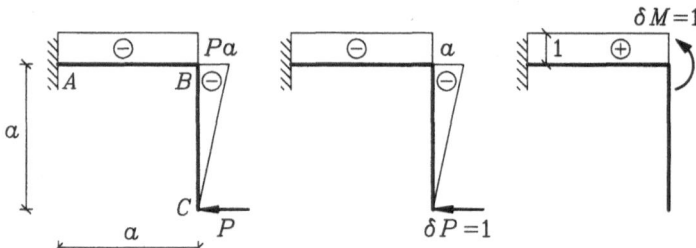

Figure 2.18: Angle beam loaded by concentrated force. a) actual moment distribution, b) virtual moment field from $\delta P = 1$ at C, c) virtual moment field from $\delta M = 1$ at B.

The left part of the figure shows the actual moment distribution $M(s)$ in the structure, while the two next figures show the virtual moment distributions $\delta M_P(s)$ and $\delta M_M(s)$ from a concentrated virtual force δP at C and a concentrated virtual moment δM at B, respectively.

The horizontal displacement w_C in the direction of the force follows from the complementary work equation as

$$w_C = \int_{ABC} \delta M_P \, \kappa \, \mathrm{d}s = \frac{1}{3} a \frac{Pa}{EI} a + a \frac{Pa}{EI} a = \frac{4Pa^3}{3EI}$$

where the integral is evaluated by the integration formula in Appendix B. Similarly the rotation θ_B is determined as

$$\theta_B = \int_{AB} \delta M_M \, \kappa \, \mathrm{d}s = -a \frac{Pa}{EI} 1 = -\frac{Pa^2}{EI}$$

Note, that this rotation contributes 75 pct. of the displacement w_C.

An important application of the complementary virtual work technique for calculation of displacements under imposed loads is the evaluation of the stiffness of a structure or structural part. This plays an important role in the analysis of statically indeterminate structures, discussed in the next section.

2.5 Statically indeterminate beams

Statically indeterminate structures are characterized by the fact that the internal forces can not be completely determined by statics alone. This property is probably most easily understood by considering a similar statically determinate structure and then adding extra supports. The extra supports generally add to the stiffness of the structure, but also make it statically indeterminate. This way of looking at statically indeterminate structures - as an equivalent statically determinate structure with extra supports added - also suggests a general technique for analyzing elastic statically indeterminate structures. The idea is to analyze the equivalent statically determinate structure while considering the reactions in the 'extra' supports as loads, that just happens to be unknown at the beginning of the analysis. This approach, called the *Force Method*, is first illustrated by a very simple example, where the steps of the method are identified. A formal statement of the method is then given in Box 2.7, and further more general examples are provided.

Example 2.11

Figure 2.19 shows a uniformly loaded beam with three simple supports. If there had been only two simple supports, the beam would have been statically determinate. This suggests imagining the same beam with one of the supports removed and replaced by an unknown force X_1, representing the reaction. This is illustrated in Fig. 2.20, where a statically equivalent beam has been obtained by removing the center support. If the center support was just removed, the beam would sag as illustrated in Fig. 2.20a. The sag at the center is counteracted by the reaction force X_1, providing an upward displacement as shown in Fig. 2.20b.

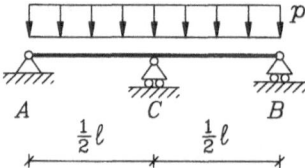

Figure 2.19: Uniformly loaded beam with three simple supports.

In the original statically indeterminate structure there is no vertical displacement of the center C of the beam due to the support. Thus the force X_1 must have a magnitude that exactly counteracts the sag of the beam without center support in Fig. 2.20a. A similar argument is used in the general case, and it is therefore convenient to introduce concepts and notation that are easily generalized. A subscript 0 is used to denote displacements generated by the original external load, acting on the statically equivalent structure. Thus, δ_{10} denotes displacement component No. 1 from the external load. The present beam is only one degree statically indeterminate, and therefore there is only one displacement component, the vertical displacement of the center of the beam, but it is convenient to use the general notation. The displacement δ_{10} is considered positive in the direction of X_1.

In principle there is no particular requirement regarding the method used to calculate the displacement δ_{10}. However, the standard procedure is to use the principle of complementary virtual work, as this leads to a very systematic method. The principle of complementary virtual work makes use of a virtual moment field generated by a concentrated unit load. This is the moment field corresponding to $X_1 = 1$, and is shown in the lower part of Fig. 2.20b. The curvature field of an elastic beam is obtained

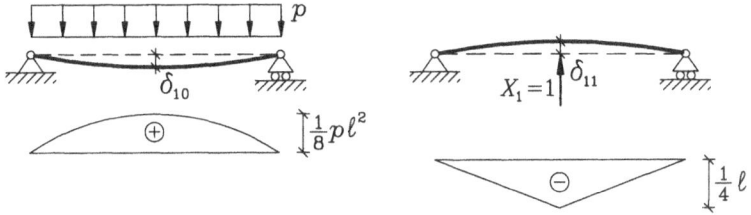

Figure 2.20: Equivalent statically determinate beam with two load
cases: a) external load $p(x)$, b) force $X_1 = 1$ at center.

as $\kappa_0(x) = M_0(x)/EI$, where M_0 is the moment field corresponding to
the external load, shown in the lower part of Fig. 2.20a. Thus, the
displacement δ_{10} is determined as

$$\delta_{10} = \int_0^\ell \frac{M_1(x)M_0(x)}{EI}\,\mathrm{d}x = -\frac{5}{384}\frac{p\,\ell^4}{EI}$$

This result was already obtained in Example 2.7. The minus is because
the positive orientation follows X_1, here selected positive upwards.

The center displacement produced by the force X_1 depends on the magnitude of X_1 and on the flexibility of the beam. The beam flexibility δ_{11}
is the displacement produced by a unit load $X_1 = 1$. Its magnitude is
found by using the moment field $M_1(x)$ corresponding to $X_1 = 1$ both
as the virtual and the actual moment field in the principle of complementary virtual work. Thus, the beam flexibility coefficient is

$$\delta_{11} = \int_0^\ell \frac{M_1(x)M_1(x)}{EI}\,\mathrm{d}x = 2\frac{1}{3}\frac{\ell}{2}\left(\frac{\ell}{4}\right)^2 = \frac{1}{48}\frac{\ell^3}{EI}$$

where the integration formula for two triangles from Appendix B has
been used.

The total displacement δ_1 at the center is obtained by superposition of
the contribution δ_{10} from the external load and the contribution $\delta_{11}X_1$
from the yet unknown force X_1. In the real structure the total displacement at the center vanishes due to the support, and thus X_1 is
determined from the displacement condition

$$\delta_1 = \delta_{10} + \delta_{11}X_1 = 0$$

whereby

$$X_1 = -\frac{\delta_{10}}{\delta_{11}} = \tfrac{5}{8}p\,\ell$$

Thus, the center support carries a little more than half the load. This could have been seen qualitatively by introducing a hinge at the beam instead of removing the center support as discussed in Exercise 2.8.

The complete solution of the statically indeterminate beam problem, consisting of reactions and internal force and displacement distributions, is obtained by superposition of the external load case of Fig. 2.20a and the load case of Fig. 2.20b, multiplied by $X_1 = \frac{5}{8}p\ell^2$. It is seen directly, that $R_C = X_1 = \frac{5}{8}p\ell^2$ and $R_A = R_B = \frac{3}{16}p\ell^2$.

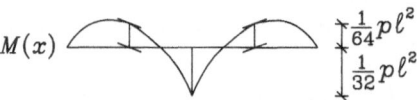

Figure 2.21: Moment distribution in the beam of Fig. 2.19.

The internal moment distribution of the statically indeterminate beam is found by combination of the the moment fields

$$M_0(x) = \tfrac{1}{2}p\,x(\ell - x)$$

and

$$M_1(x) = \begin{cases} -\tfrac{1}{2}x & \text{for } x < \tfrac{1}{2}\ell \\ -\tfrac{1}{2}(\ell - x) & \text{for } x > \tfrac{1}{2}\ell \end{cases}$$

The actual moment field is given by the linear combination,

$$M(x) = M_0(x) + M_1(x)X_1 = \begin{cases} \tfrac{1}{16}p\,x(3\ell - 8x) & \text{for } x < \tfrac{1}{2}\ell \\ \tfrac{1}{16}p(\ell - x)(8x - 5\ell) & \text{for } x > \tfrac{1}{2}\ell \end{cases}$$

and shown in Fig. 2.21. Similar relations hold for the other fields. Note, that the largest positive moment $M_{max} = \frac{9}{512}p\ell^2$ occurs at $Q(x_{max}) = R_A - p\,x_{max} = 0$, i.e. at $x_{max} = \frac{3}{16}\ell$.

Now, the general procedure of the so-called Force Method for analysis of statically indeterminate beam and frame structures can be described. A summary of the following description is given in Box 2.7.

A beam or frame structure that is statically indeterminate of degree n can be made statically determinate by releasing n constraints in the structure. This may be done by replacing supports with the corresponding forces as in Example 2.11, or by introducing hinges and applying moment couples to represent the internal moment at the location of the hinge, see e.g. the figure in Box 2.7.

In the force method a structure that is statically indeterminate of degree n is represented by an equivalent static structure, obtained by releasing n constraints and introducing imposed forces or moments $X_1, \cdots, X_n$ corresponding to the force or moment in the original constraint. In the equivalent statically determinate structure a displacement or rotation discontinuity may occur due to the release of the constraint. These discontinuities are denoted $\delta_1, \cdots, \delta_n$. The displacement discontinuities receive contributions from the external load as well as from each of the imposed loads $X_1, \cdots, X_n$. In a linear elastic structure these contributions can be combined by superposition. Thus the displacement discontinuity component δ_i can be expressed as

$$\delta_i \;=\; \delta_{i0} + \delta_{i1}X_1 + \cdots + \delta_{in}X_n \quad, \quad i = 1, \cdots, n \qquad (2.40)$$

where the term δ_{i0} is from the external load, and the coefficients δ_{ij} are the contribution to δ_i from an imposed unit load $X_j = 1$. The terms δ_{jj} represent the flexibility of the structure with respect to each of the loads loads X_j, while the terms δ_{ij} for $i \neq j$ represent the displacement discontinuity introduced by the load $X_j = 1$ at a different location.

In the real structure the constraints prevent any displacement discontinuities, and thus the imposed loads $X_1, \cdots, X_n$ are determined from the requirement, that all displacement discontinuities must vanish.

$$
\begin{aligned}
\delta_1 &= \delta_{10} + \delta_{11}X_1 + \cdots + \delta_{1n}X_n &= 0 \\
\delta_2 &= \delta_{20} + \delta_{21}X_1 + \cdots + \delta_{2n}X_n &= 0 \\
\vdots &\qquad\qquad\; \vdots &\; \vdots \\
\delta_n &= \delta_{n0} + \delta_{n1}X_1 + \cdots + \delta_{nn}X_n &= 0
\end{aligned}
\qquad (2.41)
$$

This is a linear system of n equations from which the n unknown imposed loads $X_1, \cdots, X_n$ can be determined.

The force method takes a particularly systematic form, when the flexibility coefficients δ_{i0} and δ_{ij} are determined from the principle of complementary virtual work. The term δ_{i0} is the displacement discontinuity at i generated by the external load. According to the principle of complementary work, summarized in Box 2.6, this displacement discontinuity can be found as an integral of the product of a virtual moment field $\delta M(s) = M_i(s)$ corresponding to a virtual unit load $X_i = 1$ and the curvature field $\kappa_0(s) = M_0(s)/EI$

from the external load. Thus, the terms δ_{i0} can be determined by the integrals

$$\delta_{i0} = \int M_i(s) M_0(s) \frac{ds}{EI} \quad , \quad i = 1, \cdots, n \tag{2.42}$$

where $M_0(s)$ is the moment field in the equivalent structure corresponding to the external load, and $M_i(s)$ is the moment field in the equivalent structure corresponding to an imposed unit load $X_i = 1$. The coefficients δ_{ij} are determined in a similar way from the virtual moment field $M_i(s)$ and the curvature field $\kappa_j(s) = M_j(s)/EI$ corresponding to an imposed unit load $X_j = 1$.

$$\delta_{ij} = \int M_i(s) M_j(s) \frac{ds}{EI} \quad , \quad i, j = 1, \cdots, n \tag{2.43}$$

where the moment fields correspond to the imposed unit loads $X_i = 1$ and $X_j = 1$, respectively.

In addition to providing convenient formulas for the actual computation of the coefficients δ_{i0} and δ_{ij}, the complementary virtual work principle also demonstrates that the coefficients δ_{ij} satisfy the symmetry relation

$$\delta_{ji} = \delta_{ij} \quad , \quad i, j = 1, \cdots, n \tag{2.44}$$

Thus, the equation system (2.41) for the unknown imposed loads $X_1, \cdots, X_n$ is symmetric. This important property of the method derives from the fact that kinematic variables δ_i describing the displacement discontinuities and the static variables X_i describing the corresponding loads are so-called conjugate variables. This means, that that the work of the imposed loads X_i through a small increment $d\delta_i$ of the kinematic variables is described directly by the sum

$$dW = X_1 d\delta_1 + X_2 d\delta_2 + \cdots + X_n d\delta_n \tag{2.45}$$

If a process is considered in which the imposed loads $X_1, \cdots, X_n$ are changed, the displacement discontinuities $\delta_1, \cdots, \delta_n$ will change according to the relations (2.41), whereby

$$d\delta_i = \delta_{i1} dX_1 + \cdots + \delta_{in} dX_n \quad , \quad i = 1, \cdots, n \tag{2.46}$$

If these relations are substituted into (2.45), it is seen that in the case of symmetric coefficients δ_{ij}, dW is the differential of the energy function

$$W(X_j) = \sum_i \sum_j \tfrac{1}{2} X_i \, \delta_{ij} \, X_j \tag{2.47}$$

For elastic structures the existence of an energy function $W(X_j)$ of the form (2.47) is the physical reason for the symmetry of the equations (2.41).

Box 2.7: Consistent deformations - Force Method

A statically indeterminate structure is made statically determinate by releasing n constraints, e.g. by removing supports or introducing hinges, thereby permitting displacement (angle) discontinuities $\delta_i, i = 1, \cdots, n$. External loads X_j are applied at the released constraints.

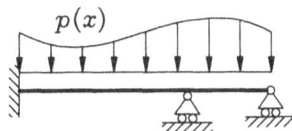

Superposition gives the total displacement (moment) discontinuity as

$$\delta_i = \delta_{i0} + \delta_{i1}X_1 + \cdots + \delta_{in}X_n \quad , \quad i = 1, \cdots, n$$

where δ_{i0} is from the external load, and the coefficients δ_{ij} is the contribution at i from a unit load $X_j = 1$.

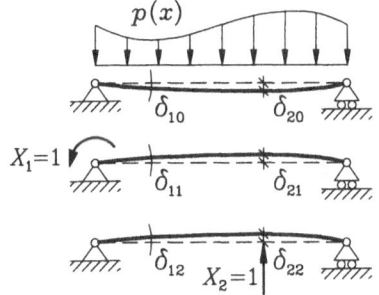

The correct $X_1, \cdots, X_n$ must cancel the displacement discontinuities,

$$
\begin{aligned}
\delta_1 &= \delta_{10} + \delta_{11}X_1 + \cdots + \delta_{1n}X_n &= 0 \\
\delta_2 &= \delta_{20} + \delta_{21}X_1 + \cdots + \delta_{2n}X_n &= 0 \\
&\ \vdots & \vdots \\
\delta_n &= \delta_{n0} + \delta_{n1}X_1 + \cdots + \delta_{nn}X_n &= 0
\end{aligned}
$$

Displacements, internal forces and reactions follow from superposition,

$$M(x) = M_0(x) + M_1(x)X_1 + \cdots + M_n(x)X_n$$

with $M_0(x), M_1(x), \cdots, M_n(x)$ from the individual load cases.

In the present formulation the static variables $X_1, \cdots, X_n$ are the unknowns of the problem. An alternative formulation, also using conjugate kinematic and static variables, but with the kinematic variables as the unknowns, is described later in this section under the name of the Deformation Method.

When the imposed loads $X_1, \cdots, X_n$ have been determined, the reactions, internal force distributions $Q(s), M(s)$ and displacement fields $w(s), \theta(s)$ of the original structure follow by superposition of the results in the $n+1$ load cases of the equivalent structure. Thus, the moment distribution in the original structure is

$$M(s) \;=\; M_0(s) + M_1(s)X_1 + \cdots + M_n(s)X_n \qquad (2.48)$$

where $M_0(s)$ is the internal moment distribution in the equivalent structure from the external load, and $M_j(s)$ is the internal moment distribution corresponding to an imposed unit load $X_j = 1$. The displacement field $w(s)$ is obtained similarly from the displacement fields $w_0(s)$ and $w_j(s)$.

Example 2.12

Figure 2.22 shows a beam of length ℓ and bending stiffness EI. At the supports the beam is connected to its surroundings in such a way that there is an elastic resistance to rotation. This is represented by the rotation spring stiffness k_A at A and k_B at B. The rotation stiffness may represent the effect of a continuously connected side span, the properties of which then determines k_A or k_B. In this example the general solution is derived, and special cases are discussed in the exercises.

The problem is solved by the force method described in Box 2.7. The idea is to consider M_A and M_B as unknowns to be determined from the conditions that the rotation θ_A and θ_B of the two end points of the beam must match the rotation of the springs. Counterclockwise rotation is considered positive.

The end rotations θ_{A0} and θ_{B0} generated by the external load, when placed on the simply supported beam without end springs, are assumed to be known from an independent analysis, e.g. by use of the principle of complementary virtual work described in Section 2.4. The rotations in the beam generated by the moments M_A and M_B are described by the coefficients $\theta_{AA}, \theta_{AB}, \theta_{BA}$ and θ_{BB}. The first subscript identifies the location of the rotation, and the second identifies the moment. Thus, θ_{AB} denotes the rotation at A generated by a unit moment acting at B.

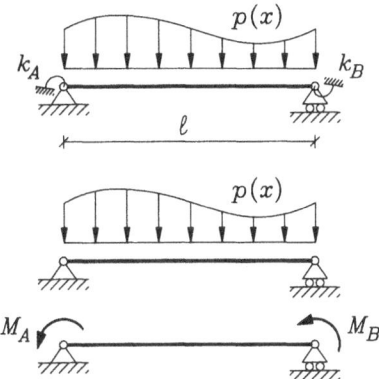

Figure 2.22: Representation of single span beam with elastic supports as a simply supported beam loaded by: a) external load $p(x)$, b) unknown end moments M_A and M_B.

In this notation the rotation at the two beam ends are

$$\theta_A = \theta_{A0} + \theta_{AA}M_A + \theta_{AB}M_B$$
$$\theta_B = \theta_{B0} + \theta_{BA}M_A + \theta_{BB}M_B$$

With the sign convention chosen here the rotation coefficients follow from Example 2.3 as

$$\theta_{AA} = \theta_{BB} = \frac{\ell}{3EI} \quad , \quad \theta_{AB} = \theta_{BA} = -\frac{\ell}{6EI}$$

When the beam is acted on by the moments M_A and M_B, the springs must be acted on by the moments $-M_A$ and $-M_B$ by the law of action and reaction. Thus, the rotation of the springs is described by

$$\theta_A = -\frac{1}{k_A} M_A \quad , \quad \theta_B = -\frac{1}{k_B} M_B$$

The angles θ_A and θ_B must be the same, whether determined from the beam or from the spring. Thus, the following equations are obtained

$$\theta_{A0} + \theta_{AA}M_A + \theta_{AB}M_B = -\frac{1}{k_A} M_A$$
$$\theta_{B0} + \theta_{BA}M_A + \theta_{BB}M_B = -\frac{1}{k_B} M_B$$

These equations are conveniently organized in the form

$$
\begin{bmatrix} 2\left(1+\dfrac{3EI}{\ell\, k_A}\right) & -1 \\[2ex] -1 & 2\left(1+\dfrac{3EI}{\ell\, k_B}\right) \end{bmatrix} \begin{bmatrix} M_A \\[2ex] M_B \end{bmatrix} = -\dfrac{6EI}{\ell} \begin{bmatrix} \theta_{A0} \\[2ex] \theta_{B0} \end{bmatrix}
$$

The solution to this system of equations is

$$
\begin{bmatrix} M_A \\[2ex] M_B \end{bmatrix} = \dfrac{-\dfrac{6EI}{\ell}}{4\left(1+\dfrac{3EI}{\ell\, k_A}\right)\left(1+\dfrac{3EI}{\ell\, k_B}\right)-1} \begin{bmatrix} 2\left(1+\dfrac{3EI}{\ell\, k_B}\right) & 1 \\[2ex] 1 & 2\left(1+\dfrac{3EI}{\ell\, k_A}\right) \end{bmatrix} \begin{bmatrix} \theta_{A0} \\[2ex] \theta_{B0} \end{bmatrix}
$$

The general solution is rather lengthy, because it contains no assumptions regarding symmetry of the load or any particular ratio between the spring constants k_A and k_B. There are three particular cases, for which the solution is considerably simpler: the fully symmetric case, symmetric beam with anti-symmetric load, and the case of only one spring.

In the fully symmetric case the springs are identical, $k_A = k_B$, and the load is symmetric, whereby $\theta_{A0} = -\theta_{B0}$. In this case the solution reduces to

$$
M_B = \dfrac{-\theta_{B0}}{\dfrac{\ell}{2EI}+\dfrac{1}{k_B}} \quad , \qquad M_A = -M_B
$$

In particular for $k_A = k_B = 0$ the moments M_A and M_B vanish, corresponding to a beam with simple supports, while for infinite stiffness the moment $M_A = -M_B = (2EI/\ell)\theta_{B0}$, corresponding to a symmetric beam with fixed ends.

For symmetric beam with anti-symmetric load $k_A = k_B$ and $\theta_{A0} = \theta_{B0}$. In this case the solution is

$$
M_B = \dfrac{-\theta_{B0}}{\dfrac{\ell}{6EI}+\dfrac{1}{k_B}} \quad , \qquad M_A = M_B
$$

For infinite stiffness the moment $M_A = M_B = -(6EI/\ell)\theta_{B0}$, corresponding to a symmetric beam with fixed ends.

If there is no rotation spring at A, the corresponding stiffness and moment vanish, i.e. $k_A = 0$ and $M_A = 0$. When taking the limit $k_A \to 0$, the term $3EI/\ell k_A$ goes to infinity, and the solution is obtained as

$$
M_B = \dfrac{-\theta_{B0}}{\dfrac{\ell}{3EI}+\dfrac{1}{k_B}} \quad , \qquad M_A = 0
$$

This is a relation similar to the symmetric and anti-symmetric cases, but with the beam stiffness represented via $3EI$.

The effect of the support is seen to be a reduction of the rotation θ_B in the full solution compared with the rotation θ_{B0} in the absence of the rotation spring. Using the spring stiffness relation it is seen that in the last case of only one rotation spring, located at B, the actual rotation is

$$\theta_B = \frac{\theta_{B0}}{1 + \dfrac{\ell\, k_B}{3EI}} \qquad , \qquad k_A = 0$$

i.e. a reduction determined by the the *relative stiffness* of the spring $\ell k_B / 3EI$. Similar relations hold in the other cases.

Example 2.13

Figure 2.23 shows a semi-infinite beam with equally spaced supports of distance ℓ. The beam is loaded by a concentrated moment M_0 at the end. This leads to a rotation θ_0 in the direction of the moment. The objective is to determine the rotation spring stiffness k_∞ in the relation

$$M_0 = k_\infty \theta_0$$

and the internal moments M_j over each of the supports, $j = 1, 2, \ldots$. The rotation stiffness of a single span beam with simple supports has been determined in Example 2.3 as $k_1 = 3EI/\ell$. The rotation stiffness at the end of a single span beam with the far end fixed was determined in Exercise 2.3 as $k_f = 4EI/\ell$. From physical reasoning the rotation stiffness k_∞ of the semi-infinite beam is expected to be given by an expression of the same form with a numerical factor between 3 and 4.

The solution is obtained by the force method. First an equivalent system is obtained by introducing a hinge at support No. 1. The solution is obtained as the sum of two load cases on the equivalent system shown in Fig. 2.24: the external load M_0, and the load by a moment couple M_1 at the support No. 1 representing the internal moment.

In the equivalent system the external load M_0 leads to a rotation θ_{00} at support No. 0 and a rotation θ_{10} of the first span at support No. 1. It follows from the results in Example 2.3 (or A.4.1 in Appendix A) that

$$\theta_{00} = \frac{\ell\, M_0}{3EI} \qquad , \qquad \theta_{10} = \frac{\ell\, M_0}{6EI}$$

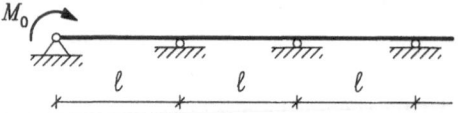

Figure 2.23: Semi-infinite beam with equally spaced simple supports.

with positive sign as shown in Fig. 2.24. In the equivalent system a discontinuity of slope occurs over support No. 1. The part of the beam to the right of support No. 1 is unloaded, and therefore the slope discontinuity from the external load is simply $\Delta\theta_{10} = \theta_{10}$.

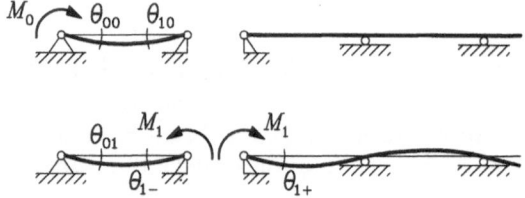

Figure 2.24: Representation of semi-infinite beam as equivalent system with: a) external load M_0, b) unknown internal moment M_1.

The discontinuity in slope is removed by introducing a moment couple M_1, shown in Fig. 2.24b. The moment M_1 is initially unknown, and therefore the effect of a unit moment couple $M_1 = 1$ is evaluated first.

In the first span a unit moment couple causes a rotation θ_{01} at support No. 0 and θ_{1-} at support No. 1.

$$\theta_{01} = \frac{\ell}{6EI} \quad , \quad \theta_{1-} = \frac{\ell}{3EI}$$

The beam to the right of support No. 1 is identical to the original beam, as the removal of a section from a semi-infinite beam does not change its properties. Thus, the rotation of the end of this semi-infinite beam under a unit moment is

$$\theta_{1+} = \frac{1}{k_\infty}$$

where the rotation stiffness k_∞ still has to be determined. The total slope discontinuity caused by a unit force couple is the sum of the contributions from left and right,

$$\Delta\theta_{11} = \theta_{1-} + \theta_{1+} = \frac{\ell}{3EI} + \frac{1}{k_\infty}$$

The magnitude of the moment couple M_1, corresponding to the internal moment at support No. 1, is determined by the condition that the slope is continuous over the support in the original structure. This is expressed by the equation

$$\Delta\theta_{10} + \Delta\theta_{11} M_1 = 0$$

from which

$$M_1 = -\frac{\Delta\theta_{10}}{\Delta\theta_{11}} = -\frac{\frac{1}{2}M_0}{1 + \dfrac{3EI}{\ell k_\infty}}$$

If k_∞ were known, this would be the solution to the problem. Observe, for instance, that removal of the second and following spans would correspond to $k_\infty = 0$ and thereby $M_1 = 0$, and at the other extreme a rotation constraint over support No. 1 would correspond to $k_\infty = \infty$ and thereby $M_1 = -\frac{1}{2}M_0$. These two cases correspond to the two single span cases mentioned at the beginning of this example.

The final condition needed to determine the value k_∞ is its original definition as the rotation stiffness at the end support of the original problem. Thus the rotation over the end support must be determined from the solution.

$$\theta_0 = \theta_{00} + \theta_{01} M_1$$

Substitution of θ_{00} and θ_{01} from above gives

$$\theta_0 = \frac{1}{4}\left(4 - \frac{1}{1 + \dfrac{3EI}{\ell k_\infty}}\right)\frac{\ell M_0}{3EI}$$

The alternative expression $\theta_0 = k_\infty^{-1} M_0$ is available from the original definition of k_∞ as the rotation stiffness at the end of the semi-infinite beam. By equating these two expressions, it is found that

$$\left(\frac{3EI}{\ell k_\infty}\right)^2 = \frac{3}{4}$$

from which

$$k_\infty = 2\sqrt{3}\frac{EI}{\ell} = 3.464\frac{EI}{\ell}$$

Indeed, this is a relation of the type conjectured at the beginning, and with a numerical factor between the value 3 for a rotation free end and the value 4 for a rotation constrained end.

It is now straightforward to complete the solution by evaluating the moment M_1,

$$M_1 \;=\; -\frac{\frac{1}{2}M_0}{1 + \dfrac{3EI}{\ell\,k_\infty}} \;=\; -\frac{M_0}{2 + \sqrt{3}} \;=\; -0.289\,M_0$$

It is clear from the equivalent system in Fig. 2.24b that the internal moment M_2 over support No. 2 can be evaluated from M_1 in precisely the same way, i.e.

$$M_2 \;=\; -\frac{M_1}{2 + \sqrt{3}} \;=\; \Big(\frac{-1}{2 + \sqrt{3}}\Big)^2 M_0$$

By continuation of this process it is seen that the internal moment M_j over support No. j is

$$M_j \;=\; \Big(\frac{-1}{2 + \sqrt{3}}\Big)^j M_0 \;=\; (-0.289)^j\,M_0$$

Note, that the internal moments M_j form a geometric sequence with negative quotient numerically less than one. Thus the moment changes sign in each span, and the magnitude decreases with distance from the end. A more detailed analysis of the solution, including shear forces and reactions, is suggested in Exercise 2.11.

Example 2.14

Roof structures often involve beams joined at an angle as shown in Fig. 2.25. The span is $2a$ and the height in the undeformed state is b. Each of the beams have length $\ell = \sqrt{a^2 + b^2}$ and bending stiffness EI.

The structure is one degree statically indeterminate. In fact, the horizontal component of the reactions can not be determined from statics alone. The internal forces are determined by the Force Method described in Box. 2.7. First a statically determinate system is created by introducing a hinge at the center as shown in Fig. 2.26. The moment distribution due to the external load can be determined directly by the following argument. The present beam bending theory disregards any change of

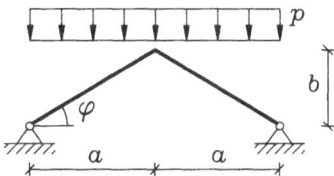

Figure 2.25: Angle beam with uniformly distributed load.

distance between the end points of a beam. Therefore the location of the center hinge is effectively fixed in space. The load intensity is p per unit horizontal length, and therefore $p\cos\varphi$ per unit length along the beam. The direction of the load is inclined φ with respect to the normal of the beam. The moment in the hinged beam is solely due to the transverse component of the load, $p\cos^2\varphi$. Thus the moment distribution due to the external distributed load is parabolic with a maximum value at the center of each beam of magnitude

$$M_{max} \;=\; \tfrac{1}{8}p\,\cos^2\varphi\,\ell^2 \;=\; \tfrac{1}{8}p\,a^2$$

The moment couple X_1, introduced at the center hinge to restore continuity, leads to triangular internal moment distributions $M_1(s)$ as shown in Fig. 2.26b.

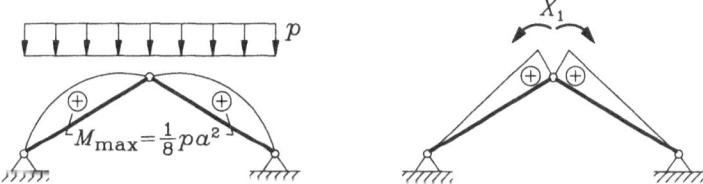

Figure 2.26: Statically determinate system, a) external load, b) center moment.

The angle discontinuity θ_1 at the center, created by the external load, is found by the equation of complementary work,

$$\theta_{10} \;=\; 2\int_0^\ell M_1(s)M_0(s)\frac{ds}{EI} \;=\; 2\,\frac{\ell}{3}\,\frac{pa^2}{8EI}\,1 \;=\; \frac{pa^2\ell}{12\,EI}$$

Similarly a unit moment $X_1 = 1$ gives the angle discontinuity

$$\theta_{11} \;=\; 2\int_0^\ell M_1(s)M_1(s)\frac{ds}{EI} \;=\; 2\,\frac{\ell}{3}\,\frac{1}{EI}\,1 \;=\; \frac{2\,\ell}{3\,EI}$$

The equation for angle continuity is

$$\theta_1 \; = \; \theta_{10} \; + \; \theta_{11}X_1 \; = \; 0$$

whereby the moment X_1 is determined as

$$X_1 \; = \; -\frac{\theta_{10}}{\theta_{11}} \; = \; -\tfrac{1}{8}pa^2$$

The moment is negative, as it contributes towards limiting the sag of the two beams in the hinged configuration.

The moment $X_1 = -\tfrac{1}{8}pa^2$ is the internal moment at the center. It lifts the parabolic moment curves in each of the two individual beams. The moment in each of these beams is the sum of the moment $M_0(s)$ from the external load, acting on the statically determinate system, and that from the moment pair X_1, acting at the central hinge. In the center of the individual beams the total moment is

$$M(\tfrac{1}{2}\ell) \; = \; M_0(\tfrac{1}{2}\ell) \; + \; M_1(\tfrac{1}{2}\ell)X_1 \; = \; = \; \tfrac{1}{8}pa^2 \; + \; \tfrac{1}{2}(-\tfrac{1}{8}pa^2) \; = \; \tfrac{1}{16}pa^2$$

Thus, the moment at $s = \tfrac{1}{2}\ell$ is reduced to half the value in the statically determinate inclined beam system. The internal moment distribution of the original statically indeterminate angle beam from Fig. 2.25 is shown in Fig. 2.27a.

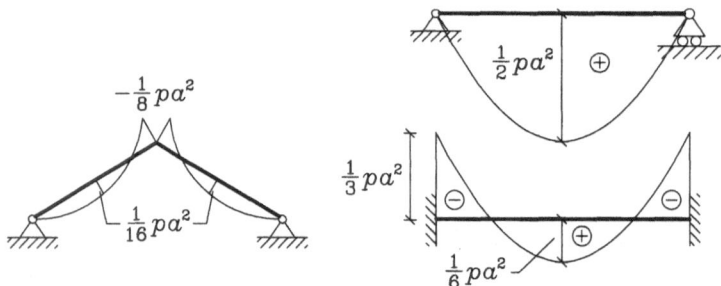

Figure 2.27: Moment distributions, a) angle beam, and a straight beam over the same span with b) simple supports, c) fixed ends.

Figure 2.27b and c shows the moment distribution in a straight beam of similar span $2a$ with simple supports and fixed ends, respectively. In both cases the internal moments are considerably larger. The moment capacity of a beam is often the deciding factor in the strength of the beam. Thus, it seems that the introduction of an angle between the beams, however small, will reduce the moment and thereby increase the load capacity considerably. Is this a miracle or an error?

The explanation is that in the present beam theory the distance between the ends of a beam is assumed to remain unaffected by the deformation. In the present problem there are two sources of axial deformation: the axial strain introduced via the normal force in the beam, and a shortening due to the curvature of the deformed shape. For a shallow angle beam the shortening due to these two effects will make the downward displacement of the center joint important, and remove the apparent stiffening from arbitrarily small angles. A beam-column theory that includes the shortening effect of axial strain and bending curvature has been developed e.g. by Krenk et al. (1999).

The Examples 2.11-2.14 have illustrated applications of the force method. The idea is to release a sufficient number of constraints in the structure to make it statically determinate. The problem can also be approached in precisely the opposite way by first introducing so many additional constraints on the structure, that its individual parts can be analyzed, and then computing the effect of releasing the imposed constraints. Beam and frame structures are constrained by preventing the motion of selected points, the *nodes*. The nodes are usually selected at the ends of beams and at joints between different beams, but in connection with approximate analysis of columns or beams on elastic foundation additional nodes may be introduced to divide the structure into smaller parts. In the special case of beam bending, described in this section, the method is called the *Deformation Method*. However, this approach is quite general and can be extended to solids, plates, shells etc. In what is called the *Finite Element Method*. Special finite elements for beams and columns are discussed in several of the following sections.

The deformation method for analysis of beam and frame structures is first introduced by a simple example, where the main concepts are identified. The general procedure is then described and summarized in Box 2.8.

Example 2.15

Figure 2.28a shows a frame consisting of two beams AC and CB of length a, rigidly joined at a right angle. The beam AC carries a transverse distributed load of intensity p.

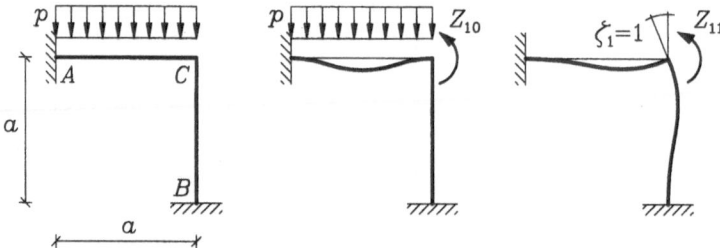

Figure 2.28: Simple frame with fixed ends, a) external load
and free corner joint, b) external load and constrained joint, c) unit rotation of joint.

The loading process is now considered as consisting of two parts. First
the load is applied together with an imposed moment Z_{10} at the joint C
of a magnitude that exactly prevents any rotation of the joint C. Then
the joint C is given a unit rotation $\zeta_1 = 1$ by application of a moment
Z_{11}. In the original problem the rotation of the joint is ζ_1 and there
is no imposed moment. The total imposed moment Z_1, following from
superposition, must therefore vanish,

$$Z_1 \; = \; Z_{10} + Z_{11}\zeta_1 \; = \; 0$$

This equation determines the rotation $\zeta_1 = -Z_{10}/Z_{11}$.

The term Z_{10} is the end moment of a beam with uniform transverse load
and fixed ends. This problem is treated in Exercise 2.4 and the result
given i Appendix A as A.3.2,

$$Z_{10} \; = \; -\tfrac{1}{12}\,p\,a^2$$

The coefficient Z_{11} is the moment needed to produce a unit rotation of
the joint C, when the beam ends at A and B are fixed. The moment
must rotate both the end of the beam AC and the end of the beam
BC. The problem of a single beam has been treated in Exercise 2.3,
and the solution given in Appendix A as A.4.2. Thus, by addition of the
contributions from the two beams,

$$Z_{11} \; = \; \frac{4EI}{a} + \frac{4EI}{a} = \frac{8EI}{a}$$

Note, that the result would have been obtained just as easily, if the two
beams had not had identical properties. The rotation of C then follows

as

$$\zeta_1 = -\frac{Z_{10}}{Z_{11}} = \frac{1}{96}\frac{p\,a^3}{EI}$$

As expected, the rotation is in the direction shown in Fig. 2.28c.

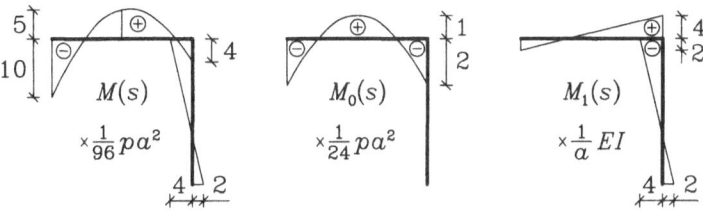

Figure 2.29: Moment distributions in frame from Fig. 2.28,
a) total moment, b) constrained joint, c) unit
joint rotation.

Reactions, internal forces and displacements of the original structure can now be obtained by superposition of the two load cases in Fig. 2.28b and c. The superposition of the internal moment is shown graphically in Fig. 2.29. Figure 2.29b shows the internal moment distribution for the case of external load and imposed constraining moment Z_{10} from Fig. 2.28b. Note, that the internal moment distribution has a discontinuity at C, exactly matching the imposed concentrated moment Z_{10}. The moment distribution corresponding to a unit rotation $\zeta_1 = 1$ is shown in Fig. 2.29c. Here, the discontinuity at C corresponds to the imposed moment Z_{11}. The internal moment distribution in the original structure of Fig. 2.28a is determined from

$$M(s) = M_0(s) + M_1(s)\zeta_1$$

and is shown in Fig. 2.29a. Note, that after superposition there is no discontinuity in the moment distribution. Also note a typical consequence of releasing the rotation constraint of C. Comparison of the moment distributions $M_0(s)$ and $M(s)$ shows that the release leads to a decrease of the moment at C, and an increase at A. Thus, the more rigid support of the beam AC attracts the larger moment.

On the basis of the example the deformation method for analysis of beam and frame structures can now be described in general terms. A number of points of the structure are designated as nodes. The initially unconstrained degrees of freedom of these nodes are denoted $\zeta_1, \cdots, \zeta_n$. These degrees of freedom may include displacements and rotations. The loading is now considered to be applied to a structure in which the motion of the nodes has been constrained by imposing forces and moments $Z_{10}, \cdots, Z_{n0}$ corresponding to the degrees of freedom $\zeta_1, \cdots, \zeta_n$. In Example 2.15 the points A, B and C are nodes, but only the rotation of C is initially unconstrained. Thus, the rotation of C is selected as ζ_1, and the corresponding Z_{10} is a moment at C. The constraining forces and moments $Z_{10}, \cdots, Z_{n0}$ are determined as the forces and moments at the end of a constrained beam, carrying the external load. In Example 2.15 the constraining moment Z_{10} is the moment at the end of the beam AC when both ends are fixed. If there had been external load on the beam BC this would have given an extra contribution to the restraining moment Z_{10}. It is interesting to reflect on the fact, that the magnitude of any restraining force or moment Z_{i0} is determined as a sum of contributions, that all come from only one beam. Thus, the relevant restraining forces of moments can in practice be collected in a small table containing the relevant load cases for a single beam. Some load cases are given in Appendix A.

The constraints are now released *one at a time* and given a unit displacement or rotation, $\zeta_j = 1$ as illustrated in the figure in Box 2.8. This unit motion $\zeta_j = 1$, with $\zeta_k = 0$ for $k \neq j$, requires that concentrated loads $Z_{1j}, \cdot, Z_{nj}$ are imposed, corresponding to the degrees of freedom $\zeta_1, \cdots, \zeta_n$. In practice, only the loads Z_{ij} associated with the beams actually being deformed will contribute, as any other beam will remain motionless without need to be constrained. Thus, in larger structures many of the coefficients Z_{ij} will vanish.

The total imposed loads Z_i are found by superposition of the contribution from the external load with constrained nodes plus a contribution from each of the the the motions $\zeta_1, \cdots, \zeta_n$,

$$Z_i = Z_{i0} + Z_{i1}\zeta_1 + \cdots + Z_{in}\zeta_n \quad , \quad i = 1, \cdots, n \qquad (2.49)$$

In the real structure there are no imposed loads, and thus the motions $\zeta_1, \cdots, \zeta_n$ must be determined by the condition that all imposed loads vanish,

$$
\begin{aligned}
Z_1 &= Z_{10} + Z_{11}\zeta_1 + \cdots + Z_{1n}\zeta_n &= 0 \\
Z_2 &= Z_{20} + Z_{21}\zeta_1 + \cdots + Z_{2n}\zeta_n &= 0 \\
\vdots & \qquad\qquad \vdots & \vdots \\
Z_n &= Z_{n0} + Z_{n1}\zeta_1 + \cdots + Z_{nn}\zeta_n &= 0
\end{aligned}
\qquad (2.50)
$$

This condition constitutes n equations for the initially unknown motions $\zeta_1, \cdots, \zeta_n$. When these motions have been determined reactions and the internal force and displacement distributions follow from superposition according to the recipe

$$M(x) = M_0(x) + M_1(x)\zeta_1 + \cdots + M_n(x)\zeta_n \qquad (2.51)$$

where $M_0(s)$ is the internal moment distribution from external loads on the constrained structure, while $M_j(s)$ is the internal moment distribution from the unit motion $\zeta_j = 1$, with $\zeta_k = 0$ for $k \neq j$.

Although in practice the constraint loads Z_{i0} and Z_{ij} are often read from a table of simple load cases for a single beam, it is instructive to review a procedure for calculation of the coefficients Z_{ij} by use of the principle of virtual work. Consider the case with all nodes constrained, except $\zeta_i = 1$. The corresponding loads are $Z_{i1}, \cdots, Z_{in}$ and the internal moment field is $M_i(s)$. This combination of load and internal moment field is now used in the principle of virtual work together with the virtual deformation field corresponding to the unit motion $\delta\zeta_j = 1$. The corresponding virtual curvature field is $\delta\kappa_j = M_j(s)/EI$. There are no distributed loads or discontinuities in the virtual displacement field, and it therefore follows from the principle of virtual work in Box 2.5 that the external work $Z_{ij}\delta\zeta_j$ is given as

$$Z_{ij} = \int M_i(s) M_j(s) \frac{\mathrm{d}s}{EI} \quad , \quad i,j = 1, \cdots, n \qquad (2.52)$$

where the moment fields correspond to the imposed isolated unit motions $\zeta_i = 1$ and $\zeta_j = 1$, respectively. The integral relation (2.52) implies, that the coefficients Z_{ij} satisfy the symmetry relations

$$Z_{ji} = Z_{ij} \quad , \quad i,j = 1, \cdots, n \qquad (2.53)$$

Thus the equation system (2.50) of the deformation method is symmetric.

In the deformation method the variables Z_i and ζ_i are conjugate forming the work increment

$$dW_c = \zeta_1 dZ_1 + \zeta_2 dZ_2 + \cdots + \zeta_n dZ_n \qquad (2.54)$$

Substitution of $dZ_1, \cdots, dZ_n$ from (2.50) and integration leads to the complementary energy function

$$W_c(\zeta_j) = \sum_i \sum_j \tfrac{1}{2}\zeta_i Z_{ij} \zeta_j \qquad (2.55)$$

where Z_{ij} appear as stiffness coefficients.

Box 2.8: Constrained nodes - Deformation Method

The nodes of the structure are fixed by imposing n constraint loads $Z_{10}, \cdots, Z_{n0}$ to the initially unconstrained degrees of freedom $\zeta_1, \cdots, \zeta_n$.

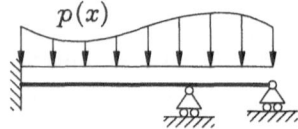

The constraints are now released *one at a time*. A unit deformation $\zeta_j = 1$ of the unloaded structure requires imposed loads $Z_{1j}, \cdots, Z_{nj}$.

The total imposed load Z_i for external load and released constraints is given by superposition as

$$Z_i = Z_{i0} + Z_{i1}\zeta_1 + \cdots + Z_{in}\zeta_n \quad , \quad i = 1, \cdots, n$$

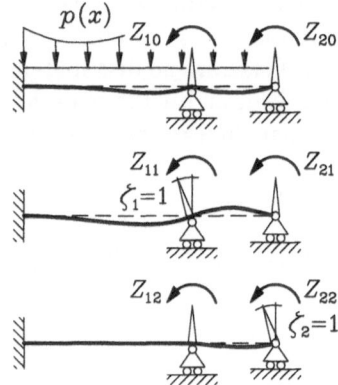

The correct $\zeta_1, \cdots, \zeta_n$ cancel the imposed constraint loads,

$$Z_1 = Z_{10} + Z_{11}\zeta_1 + \cdots + Z_{1n}\zeta_n = 0$$
$$\vdots \qquad \vdots \qquad \qquad \vdots$$
$$Z_n = Z_{n0} + Z_{n1}\zeta_1 + \cdots + Z_{nn}\zeta_n = 0$$

Displacements, internal forces and reactions follow from superposition,

$$M(x) = M_0(x) + M_1(x)\zeta_1 + \cdots + M_n(x)\zeta_n$$

with $M_0(x), M_1(x), \cdots, M_n(x)$ from the individual load cases.

Example 2.16

Figure 2.30 shows a beam ABC of constant bending stiffness EI with the left end A fixed, a simple support B at a distance a, and another simple support C at a further distance b. The load consists of a uniform transverse force distribution of intensity p. The relative lengths of the spans determine the rotation of the nodes B and C as described in the following.

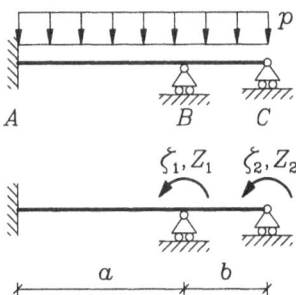

Figure 2.30: Two-span beam with uniformly distributed load. a) external load, b) rotations and constraint moments.

The nodes are A, B and C, and the degrees of freedom to be constrained are the rotations ζ_1 and ζ_2 of B and C, respectively. The moment Z_{10} at B is the sum of the moments needed to prevent rotation of the two uniformly loaded beams AB and BC. The magnitude is given in Apendix A as case A.3.2. The moment Z_{20} corresponds to constraining rotation of the end of the beam BC only.

$$Z_{10} = -\tfrac{1}{12}pa^2 + \tfrac{1}{12}pb^2 \quad , \quad Z_{20} = -\tfrac{1}{12}pb^2$$

The rotation $\zeta_1 = 1$ requires rotation of both beams, while the rotation $\zeta_2 = 1$ only requires rotation of the beam BC. The necessary moments follow from the load case A.4.2 in Appendix A. Note, that the moments Z_{ij} are positive in the counter-clockwise direction, while the appendix gives the internal moments.

$$Z_{11} = 4EI/a + 4EI/b \quad , \quad Z_{12} = 2EI/b$$
$$Z_{21} = 2EI/b \qquad\qquad , \quad Z_{22} = 4EI/b$$

The equation system (2.50) for the rotations then is

$$\frac{2EI}{ab} \begin{bmatrix} 2(a+b) & a \\ a & 2a \end{bmatrix} \begin{bmatrix} \zeta_1 \\ \zeta_2 \end{bmatrix} = \frac{p}{12} \begin{bmatrix} a^2 - b^2 \\ b^2 \end{bmatrix}$$

The solution to these equations is

$$
\begin{bmatrix} \zeta_1 \\ \zeta_2 \end{bmatrix} = \frac{p}{24EI} \frac{b}{3a+4b} \begin{bmatrix} a(2a^2 - 3b^2) \\ (a+b)^2(2b-a) \end{bmatrix}
$$

It is seen from this result that $\zeta_1 > 0$ for $a > \sqrt{3/2}\,b$, and $\zeta_2 > 0$ for $a < 2b$. This confirms the general intuition, that $\zeta_1 > 0$ for a very long span AB, while $\zeta_1 < 0$ and $\zeta_2 > 0$ for a very long span BC.

Determination of the internal moment distribution and the reactions is discussed in Exercise 2.14.

The two methods presented here for analysis of statically indeterminate beam structures are both based on the principle of superposition. However, there are two essential differences. In the force method the statically indeterminate structure is replaced by an equivalent statically determinate structure. Thus, the analysis of a full structure is still required, and although this analysis is statically determinate and may be quite simple, it is not trivially adapted for automatic computation. In contrast, the deformation method operates only on a single beam at a time. First the load Z_{i0}, necessary to constrain a node, is obtained by adding the contributions from each beam connected to that node, and similarly the loads Z_{ij} only involves nodes connected directly to the node with the degree of freedom ζ_i. Thus, the building block of the deformation method is the deformation characteristics of the individual beam. This is of great importance for the development of systematic numerical procedures for analysis of structures.

The other difference between force and deformation methods is, that in the force method the number of variables X_i is equal to the degree of static indeterminacy of the original structure. In the deformation method all nodes connected to more than one beam must be constrained in order to permit direct evaluation of the constraining loads Z_{i0} and Z_{ij} from standard load cases. This may lead to a very large number of constraints, but permits the analysis of even statically determinate structures in a matrix oriented beam by beam fashion. The simple beam bending element is considered in the next section, and more general elements for beams on elastic foundation, vibrating beams, and beam-columns are described later. These elements form the basis of most beam and frame analysis carried out in engineering practice.

2.6   The beam bending element

The deformation method, described in the previous section, relies on designating a suitable number of points of the structure as nodes, and first considering the structure with all nodes fully constrained, and then giving each degree of freedom a unit motion, one at a time. The essential component in this procedure is the stiffness of each member, when one of its nodes is given a unit motion. This procedure can be generalized to plates, shells and solids, Cook et al. (1989), and can be developed into a general mathematical tool for approximate solution of ordinary and partial differential equations, see e.g. Zienkiewicz & Taylor (1989, 1991). The present text is limited to beam structures, but even within this limited class of structures it is important to develop several different types of elements, e.g. for beams on elastic foundation, vibrating beams, and beam-columns. In this section the basic beam bending element is first derived directly from the equations of statics, and then the alternative method of virtual work is presented in a form suitable for later use.

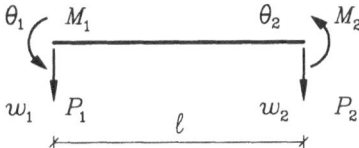

Figure 2.31: Homogeneous beam with end loads (P_1, M_1, P_2, M_2).

Figure 2.31 shows a beam loaded by transverse forces P_1, P_2 and moments M_1, M_2 at the ends. The beam is assumed to be homogeneous with bending stiffness EI. An alternative approach valid for non-homogeneous beams is discussed in the next section.

The transverse displacement $w(x)$ of the beam element is governed by the equilibrium equation (2.21),

$$\frac{d^2}{dx^2}\left(EI\,\frac{d^2w}{dx^2}\right) - p(x) = 0 \tag{2.56}$$

In the absence of distributed transverse load $p(x) = 0$, and the displacement is a third degree polynomial of the form

$$w(x) = C_0 + C_1 x + C_2 x^2 + C_3 x^3 \tag{2.57}$$

where $C_0, \cdots, C_3$ are arbitrary constants to be determined by the boundary conditions. Although this is a fully valid representation of the displacement distribution, it is not very convenient, because the arbitrary constants and the powers $1, \cdots, x^3$ do not have any direct relations to the mechanics of the problem.

Figure 2.32: Shape functions for imposed unit end motions $w_1, \theta_1, w_2, \theta_2$.

A more convenient representation of the displacement distribution is suggested by Fig. 2.32, showing the displacement distribution corresponding to each of the unit motions. These normalized displacement distributions are called *shape functions*. The complete solution corresponding to arbitrary motion of the end points follows from superposition in the form

$$
\begin{aligned}
w(x) \;=\; & \left(1+2\frac{x}{\ell}\right)\left(1-\frac{x}{\ell}\right)^2 w_1 \;-\; \frac{x}{\ell}\left(1-\frac{x}{\ell}\right)^2 \ell\,\theta_1 \\
& +\; \left(\frac{x}{\ell}\right)^2\left(3-2\frac{x}{\ell}\right)w_2 \;+\; \left(\frac{x}{\ell}\right)^2\left(1-\frac{x}{\ell}\right)\ell\,\theta_2
\end{aligned}
\tag{2.58}
$$

This representation can be constructed directly from the end conditions and is easily verified by differentiation. The shape functions, appearing as factors to $w_1, \theta_2, w_2, \theta_2$ and shown in Fig. 2.32, are denoted

$$
\begin{aligned}
N_1(x) \;&=\; \left(1+2\frac{x}{\ell}\right)\left(1-\frac{x}{\ell}\right)^2 \quad , \quad & N_2(x) \;&=\; -\frac{x}{\ell}\left(1-\frac{x}{\ell}\right)^2 \ell \\
N_3(x) \;&=\; \left(\frac{x}{\ell}\right)^2\left(3-2\frac{x}{\ell}\right) \quad , \quad & N_4(x) \;&=\; \left(\frac{x}{\ell}\right)^2\left(1-\frac{x}{\ell}\right)\ell
\end{aligned}
\tag{2.59}
$$

It is now convenient to introduce matrix notation, here denoted by sans serif type. The displacement components are collected in the column vector u, defined by its transpose

$$
\mathsf{u}^T \;=\; [\,w_1, \theta_1, w_2, \theta_2\,]
\tag{2.60}
$$

The shape functions from (2.59) are collected in the row vector

$$
\mathsf{N}(x) \;=\; [\,N_1(x), N_2(x), N_3(x), N_4(x)\,]
\tag{2.61}
$$

In this notation the displacement field $w(x)$ in (2.58) takes the compact form

$$w(x) = N_1(x)w_1 + N_2(x)\theta_1 + N_3(x)w_2 + N_4(x)\theta_2 = \mathsf{N}(x)\mathsf{u} \quad (2.62)$$

This form is standard for representation of the displacement field of structures and solids in the finite element method.

The purpose of the present analysis is to obtain the forces and moments at the ends of the beam. They follow from the internal force and moment distributions, given in Box 2.4 as derivatives of the displacement field. For a Bernoulli-beam element with constant bending stiffness

$$M(x) = -EI\,\mathrm{d}^2w(x)/\mathrm{d}x^2 = -EI\,[\mathrm{d}^2\mathsf{N}(x)/\mathrm{d}x^2]\,\mathsf{u}$$

$$\qquad\qquad\qquad\qquad\qquad\qquad\qquad\qquad\qquad (2.63)$$

$$Q(x) = -EI\,\mathrm{d}^3w(x)/\mathrm{d}x^3 = -EI\,[\mathrm{d}^3\mathsf{N}(x)/\mathrm{d}x^3]\,\mathsf{u}$$

Thus the introduction of the matrix notation leads to a product form of the internal moment and shear force quite similar to (2.62) for the displacement field.

The final step is to collect the forces and moments at the beam ends in the column vector f, defined by its transpose

$$\mathsf{f}^T = [\,P_1, M_1, P_2, M_2\,] \quad (2.64)$$

This nodal force vector can be expressed in terms of the displacement vector u by reference to the internal forces at the ends of the beam and use of (2.63),

$$\begin{bmatrix} P_1 \\ M_1 \\ P_2 \\ M_2 \end{bmatrix} = \begin{bmatrix} -Q(0) \\ -M(0) \\ Q(\ell) \\ M(\ell) \end{bmatrix} = EI \begin{bmatrix} [\mathrm{d}^3\mathsf{N}(x)/\mathrm{d}x^3]_{x=0}\,\mathsf{u} \\ [\mathrm{d}^2\mathsf{N}(x)/\mathrm{d}x^2]_{x=0}\,\mathsf{u} \\ -[\mathrm{d}^3\mathsf{N}(x)/\mathrm{d}x^3]_{x=\ell}\,\mathsf{u} \\ -[\mathrm{d}^2\mathsf{N}(x)/\mathrm{d}x^2]_{x=\ell}\,\mathsf{u} \end{bmatrix} \quad (2.65)$$

This is a matrix relation in which each of the four rows contain the factor u. It can therefore be written in the compact form

$$\mathsf{f} = \mathsf{K}\,\mathsf{u} \quad (2.66)$$

where u is the displacement vector, defined in (2.60), f is the nodal force vector, defined in (2.64), and K is the beam element stiffness matrix, obtained from (2.65) as

$$\mathsf{K} = \frac{EI}{\ell^3} \begin{bmatrix} 12 & -6\ell & -12 & -6\ell \\ -6\ell & 4\ell^2 & 6\ell & 2\ell^2 \\ -12 & 6\ell & 12 & 6\ell \\ -6\ell & 2\ell^2 & 6\ell & 4\ell^2 \end{bmatrix} \quad (2.67)$$

This is the *stiffness matrix* for bending of a Bernoulli beam. Note, that the stiffness matrix is symmetric. This property follows directly, if the stiffness matrix is derived from the principle of virtual work as demonstrated below.

The stiffness matrix contains terms of different dimension, due to the different dimensions of translations and rotations, and forces and moments. Not all of the terms in the stiffness matrix are independent. In order for the element of be in equilibrium the sum of the transverse forces $P_1 + P_2$ must vanish, and this condition must hold for any combination of the displacement variables. From this it follows that the third row is obtained from the first by changing the sign of all terms. There are also relations due to moment equilibrium and the fact that rigid body motion does not lead to internal forces. This suggests that the stiffness matrix can be built up from a smaller number of terms, a possibility explored in connection with shear flexibility in the next section and for beam-columns in Section 3.2.

Example 2.17

The element format (2.66) is set up as if all nodal forces are known and the nodal displacements are unknown. However, in order to have a unique solution boundary conditions must be imposed, representing the supports. This is illustrated by considering a homogeneous elastic cantilever beam of length ℓ and bending stiffness EI, shown in Fig. 2.33. The beam is loaded at the tip with a concentrated force P and a concentrated moment M.

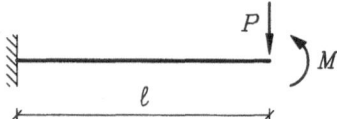

Figure 2.33: Cantilever beam with end force P and moment M.

When the beam is considered as a single beam element the nodal forces and displacements are related by (2.66). In the present case the nodal displacements w_1, θ_1 at the support and the nodal forces P_2, M_2 at the tip are known. Thus the matrix relation (2.66) takes the form

$$
\frac{EI}{\ell^3}
\begin{bmatrix}
12 & -6\ell & -12 & -6\ell \\
-6\ell & 4\ell^2 & 6\ell & 2\ell^2 \\
-12 & 6\ell & 12 & 6\ell \\
-6\ell & 2\ell^2 & 6\ell & 4\ell^2
\end{bmatrix}
\begin{bmatrix}
0 \\
0 \\
\boxed{w_2} \\
\boxed{\theta_2}
\end{bmatrix}
=
\begin{bmatrix}
\boxed{P_1} \\
\boxed{M_1} \\
P \\
M
\end{bmatrix}
$$

At the beginning of the analysis the nodal displacements w_2, θ_2 and the nodal forces P_1, M_2, indicated by boxes, are unknown.

The solution is obtained by first solving the two last equations for w_2, θ_2.

$$\frac{EI}{\ell^3} \begin{bmatrix} 12 & 6\ell \\ 6\ell & 4\ell^2 \end{bmatrix} \begin{bmatrix} w_2 \\ \theta_2 \end{bmatrix} = \begin{bmatrix} P \\ M \end{bmatrix}$$

The solution of these two linear equations is

$$\begin{bmatrix} w_2 \\ \theta_2 \end{bmatrix} = \frac{\ell}{6EI} \begin{bmatrix} 2\ell^2 & -3\ell \\ -3\ell & 6 \end{bmatrix} \begin{bmatrix} P \\ M \end{bmatrix}$$

The tip displacement $w = P\ell^3/3EI$ from the force was already found in Examples 2.8 and 2.9.

When the displacements w_2, θ_2 have been determined, the two first equations can be used to calculate the forces P_1, M_1 at the support. Substitution of the displacements give

$$\begin{bmatrix} P_1 \\ M_1 \end{bmatrix} = \frac{EI}{\ell^3} \begin{bmatrix} -12 & -6\ell \\ 6\ell & 2\ell^2 \end{bmatrix} \begin{bmatrix} w_2 \\ \theta_2 \end{bmatrix} = \begin{bmatrix} -1 & 0 \\ \ell & -1 \end{bmatrix} \begin{bmatrix} P \\ M \end{bmatrix}$$

Thus, the nodal forces at the support are

$$P_1 = -P \quad , \qquad M_1 = P\ell - M$$

as is easily established directly from statics. It is seen that in the finite element approach the unknown displacement components are first determined from a stiffness relation of reduced size, and the remaining nodal forces can then be recovered from those equations, initially left out. In this procedure no distinction is made between statically determinate or indeterminate structures.

———

An interesting alternative to direct derivation of the beam bending stiffness matrix (2.67), demonstrated above, consists in using the principle of virtual work. This approach will demonstrate the symmetry of the stiffness matrix and also produce an approximate method that can be used in more complicated problems. Only a single beam is considered, and the principle of virtual work can therefore be used in its simple form (2.26). In terms of the components of the force vector (2.64) the virtual work equation (2.26) is

$$\delta w_1 P_1 + \delta \theta_1 M_1 + \delta w_2 P_2 + \delta \theta_2 M_2 +$$

$$+ \int_0^\ell \delta w \, p(x) \, dx = \int_0^\ell \delta \kappa(x) \, M(x) \, dx \tag{2.68}$$

The beam is assumed to be elastic, and the internal moment is expressed by the beam curvature, $M(x) = EI\,\kappa = -EI\,dw^2/dx^2$. When the virtual curvature is similarly introduced as $\delta\kappa(x) = -d^2(\delta w)/dx^2$, the virtual work equation (2.68) takes the form

$$\delta w_1 P_1 + \delta\theta_1 M_1 + \delta w_2 P_2 + \delta\theta_2 M_2 +$$
$$+ \int_0^\ell \delta w\, p(x)\, dx \; = \; \int_0^\ell \frac{d^2(\delta w)}{dx^2}\, EI\, \frac{d^2 w}{dx^2}\, dx \qquad (2.69)$$

Now comes the basic assumption, namely that the displacement field $w(x)$ and the virtual displacement field $\delta w(x)$ can be represented by the shape functions $N_1(x), \cdots, N_4(x)$ given in (2.56). This corresponds to representations of the form

$$w(x) \;=\; \mathsf{N}(x)\,\mathsf{u} \quad , \qquad \delta w(x) \;=\; \mathsf{N}(x)\,\delta\mathsf{u} \;=\; \delta\mathsf{u}^T\,\mathsf{N}(x)^T \qquad (2.70)$$

where u was given in (2.60) and $\delta\mathsf{u}$ is defined similarly by

$$\delta\mathsf{u}^T \;=\; [\,\delta w_1, \delta\theta_1, \delta w_2, \delta\theta_2\,] \qquad (2.71)$$

The last equality in (2.70b) is the transpose of a scalar, where the order of the terms are interchanged and transposed, without changing the value of the product.

The element stiffness is represented by the integral on the right side of (2.69). Thus, for an unloaded beam substitution of $\delta w(x)$ and $w(x)$ from (2.70) gives

$$\delta\mathsf{u}^T\,\mathsf{f} \;=\; \delta\mathsf{u}^T\Big[\int_0^\ell EI\, \frac{d^2\mathsf{N}^T}{dx^2}\frac{d^2\mathsf{N}}{dx^2}\, dx\Big]\mathsf{u} \qquad (2.72)$$

The integral in the square brackets is a 4 by 4 matrix

$$\mathsf{K} \;=\; \int_0^\ell EI\, \frac{d^2\mathsf{N}^T}{dx^2}\frac{d^2\mathsf{N}}{dx^2}\, dx \qquad (2.73)$$

where the matrix element K_{ij} is obtained by combining shape function $N_i(x)$ with $N_j(x)$. With this notation the equation (2.72) can be written as

$$\delta\mathsf{u}^T\left(\mathsf{f} - \mathsf{K}\mathsf{u}\right) \;=\; 0 \qquad (2.74)$$

The components of the virtual displacement vector $\delta\mathsf{u}$ can be chosen arbitrarily, and thus the equation (2.74) implies that the vector in the parenthesis must vanish. Thus the virtual work equation leads to a stiffness relation of the form (2.66), but with the stiffness matrix K defined by the integral formula (2.73). Note, that the use of the same shape functions in $w(x)$ and $\delta w(x)$ leads to a symmetric stiffness matrix.

A crucial step in the present use of the principle of virtual work is the substitution of the moment field $M(x) = -EI\, dw^2/dx^2$. If an exact representation of the displacement field $w(x)$ is used, the integral formula (2.73) is exact. For a beam with constant bending stiffness EI the representation of the displacement field $w(x)$ in terms of the shape functions $N_1(x), \cdots, N_4(x)$ is exact, and the integral formula for the stiffness matrix therefore reproduces the result (2.67), already obtained by direct evaluation of the internal forces. The main interest of the virtual work approach to evaluation of stiffness properties of beams - as well as other structures - derives from the fact that this method is easily extended to produce approximate stiffness matrix expressions for problems governed by more complicated differential equations. The integral formula (2.73) is often used in practice for beams with bending stiffness variation along the beam. However, for that particular case a more accurate method is available as described in the next section.

If a load is present on the beam element an extra term enters the virtual work equation. Substitution of the shape function representation of the virtual displacement field $\delta w(x)$ gives

$$\int_0^\ell \delta w\, p(x)\ dx \;=\; \delta u^T \left[\, \int_0^\ell \mathsf{N}(x)^T\, p(x)\ dx \right] \tag{2.75}$$

The integral in the square brackets is a column vector representing the load. It is convenient to introduce this equivalent load vector as

$$\mathsf{f}_0 \;=\; \int_0^\ell \mathsf{N}(x)^T\, p(x)\ dx \tag{2.76}$$

Thus the full statement of the virtual work equation for a loaded beam element is

$$\delta u^T \left(\mathsf{f} + \mathsf{f}_0 - \mathsf{K}u \right) \;=\; 0 \tag{2.77}$$

The components of the virtual displacement vector δu can be chosen arbitrarily, leading to the vector equation

$$\mathsf{f} + \mathsf{f}_0 \;=\; \mathsf{K}\,u \tag{2.78}$$

Like the principle of virtual work states that external virtual work must equal internal virtual work, the vector equation (2.78) states that the forces on a beam element, represented by the sum of the forces at the nodes f and the external load f_0, must equal the internal forces $\mathsf{K}u$, generated by the deformation of the beam.

Box 2.9: Beam bending element - Shape functions

The stiffness matrix K of a beam element establishes a relation of the form $\mathsf{f} = \mathsf{K}\,\mathsf{u}$ between node displacement components u and corresponding nodal force components f,

$$\mathsf{u}^T = [\,w_1, \theta_1, w_2, \theta_2\,] \quad , \quad \mathsf{f}^T = [\,P_1, M_1, P_2, M_2\,]$$

For a Bernoulli beam element - see Box 2.3 - the stiffness matrix is

$$\mathsf{K} = \frac{EI}{\ell^3} \begin{bmatrix} 12 & -6\ell & -12 & -6\ell \\ -6\ell & 4\ell^2 & 6\ell & 2\ell^2 \\ -12 & 6\ell & 12 & 6\ell \\ -6\ell & 2\ell^2 & 6\ell & 4\ell^2 \end{bmatrix}$$

The stiffness matrix can be calculated by solution of the beam equilibrium equation or from the principle of virtual work as

$$\mathsf{K} = \int_0^\ell EI \, \frac{\mathrm{d}^2\mathsf{N}^T}{\mathrm{d}x^2} \frac{\mathrm{d}^2\mathsf{N}}{\mathrm{d}x^2} \, \mathrm{d}x$$

using a representation of the displacement field

$$w(x) = N_1(x)w_1 + N_2(x)\theta_1 + N_3(x)w_2 + N_4(x)\theta_2 = \mathsf{N}(x)\mathsf{u}$$

in terms of shape functions $\mathsf{N}(x) = [\,N_1(x), \cdots, N_4(x)\,]$.

Example 2.18

The load is represented by the vector

$$\mathsf{f}_0^T = [\,P_1^0, M_1^0, P_2^0, M_2^0\,]$$

The special case of a uniformly distributed load of constant intensity p is shown in Fig. 2.34.

Figure 2.34: Equivalent nodal loads $f_0^T = [\, P_1^0, M_1^0, P_2^0, M_2^0 \,]$.

The components of the nodal load vector f_0 are calculated from the integral relation (2.76). When introducing the non-dimensional length coordinate $\xi = x/\ell$ into the shape functions (2.59) the following results are obtained.

$$P_1^0 \;=\; \int_0^\ell N_1(x)\, p \, \mathrm{d}x \;=\; p\ell \int_0^1 (1+2\xi)(1-\xi)^2 \, \mathrm{d}\xi \;=\; \tfrac{1}{2}p\ell$$

$$M_1^0 \;=\; \int_0^\ell N_2(x)\, p \, \mathrm{d}x \;=\; -p\ell^2 \int_0^1 \xi(1-\xi)^2 \, \mathrm{d}\xi \;=\; -\tfrac{1}{12}p\ell^2$$

The complete nodal load vector then follows from symmetry,

$$f_0^T \;=\; [\, \tfrac{1}{2}p\ell, -\tfrac{1}{12}p\ell^2, \tfrac{1}{2}p\ell, \tfrac{1}{12}p\ell^2 \,]$$

The nodal loads in this case are seen to be identical to the reaction forces and moments of a uniformly loaded beam with fixed ends, see case A.3.2 in Appendix A.

The format (2.78) of the equilibrium equations enables a reformulation of the deformation method in matrix format. This format, in which the structure is divided into elements that each give locally defined contributions to the behaviour of the complete structure, goes by the name of the Finite Element Method. In the present context of beam structures the main difference between the finite element method and the deformation method is the procedure used for setting up the equations for the full structure. The finite element approach is briefly illustrated in the following example. However, the focus of the present is on the behaviour of the individual elements, and for a more detailed description of the finite element approach to structural analysis the literature must be consulted, e.g. Thelanderson (1984) and Cook et al. (1989).

Example 2.19

The finite element approach to beam problems is illustrated for the uniformly loaded two-span beam treated by the deformation method in Example 2.16. The problem is illustrated in Fig. 2.35.

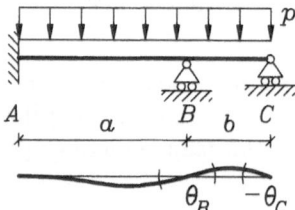

Figure 2.35: Two-span beam. a) uniformly distributed load, b) displacement degrees of freedom θ_B, θ_C.

The structure consists of two elements, AB of length a and BC of length b. Element properties are indicated with index a and b in the following. The element stiffness matrices follow directly from (2.67),

$$
\mathsf{K}_a = \frac{EI}{a^3}
\begin{bmatrix}
12 & -6a & -12 & -6a \\
-6a & 4a^2 & 6a & 2a^2 \\
-12 & 6a & 12 & 6a \\
-6a & 2a^2 & 6a & 4a^2
\end{bmatrix}
, \quad
\mathsf{K}_b = \frac{EI}{b^3}
\begin{bmatrix}
12 & -6b & -12 & -6b \\
-6b & 4b^2 & 6b & 2b^2 \\
-12 & 6b & 12 & 6b \\
-6b & 2b^2 & 6b & 4b^2
\end{bmatrix}
$$

and the nodal load vectors follow from Example 2.18,

$$
\mathsf{f}_a^{0T} = [\,\tfrac{1}{2}, -\tfrac{1}{12}a, \tfrac{1}{2}, \tfrac{1}{12}a\,]\,pa
\quad , \quad
\mathsf{f}_b^{0T} = [\,\tfrac{1}{2}, -\tfrac{1}{12}b, \tfrac{1}{2}, \tfrac{1}{12}b\,]\,pb
$$

These contributions are additive - stiffness from both elements must be included, and loads on both elements accounted for. The displacements are different in the sense that the displacement components w, θ at a node are common for all elements joined to that node. In the present example this means that the displacement vectors of the elements are

$$
\mathsf{u}_a^T = [\,w_A, \theta_A, w_B, \theta_B\,]
\quad , \quad
\mathsf{u}_b^T = [\,w_B, \theta_B, w_C, \theta_C\,]
$$

where w_B, θ_B are common for the two elements.

The superposition of contributions from the two elements is illustrated
in the following schematic formula.

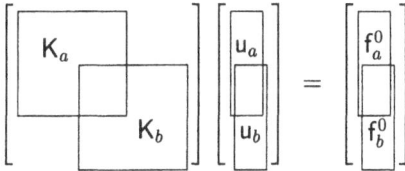

The displacement vector of the complete structure has 6 components,
w_j, θ_j for each of the nodes A, B and C. The nodal load vector of the
complete structure has a corresponding set of 6 components, P_j, M_j for
each of the nodes. The two elements are joined at node B. Thus, both
element load vectors f_a^0 and f_b^0 contribute to the load at node B by direct
addition of the contribution from each of the elements. The internal
forces $K_a u_a$ and $K_b u_b$ from both elements also contribute to the left side
of the equation as indicated. Thus, in the total 6 by 6 stiffness matrix
of the structure the central 2 by 2 block is the sum of the lower right 2
by 2 block of K_a and the upper left 2 by 2 block of K_b. This addition
procedure is easily automated by setting up a list, that identifies where
the element displacement components occur in the global displacement
vector.

In the present problem only two of the displacement components, θ_B
and θ_C, are different from zero. These components are number 4 and 6
in the global displacement vector, and therefore correspond to equations
No. 4 and 6, respectively. The non-vanishing terms of two equations can
be identified directly from the sketch above. The equations are

$$\begin{bmatrix} K_{44}^a + K_{22}^b & K_{24}^b \\ K_{42}^{rb} & K_{44}^h \end{bmatrix} \begin{bmatrix} \theta_B \\ \theta_C \end{bmatrix} = \begin{bmatrix} f_4^{0a} + f_2^{0b} \\ f_4^{0b} \end{bmatrix}$$

The values of the terms are found from the matrices given above.

$$EI \begin{bmatrix} 4a^{-1} + 4b^{-1} & 2b^{-1} \\ 2b^{-1} & 4b^{-1} \end{bmatrix} \begin{bmatrix} \theta_B \\ \theta_C \end{bmatrix} = \begin{bmatrix} \frac{1}{12}pa^2 - \frac{1}{12}pb^2 \\ \frac{1}{12}pb^2 \end{bmatrix}$$

This is exactly the equation system arrived at by the deformation method
in Example 2.16. The main difference is in the assembly of the equations
for the complete structure. In the deformation method only the effect of
unconstrained degrees of freedom are analyzed, and the coefficients are
identified by 'the effect at node i from a displacement at node j'. In the
finite element approach the stiffness matrix and load vector are set up

for all elements, and in principle the full equation system for *all* degrees of freedom is formed, and subsequently reduced for boundary conditions and solved. This indicates, that the deformation method is often most convenient for hand calculations, while the finite element approach lends itself more easily to automatic computation.

The beam bending element has been introduced in this section in its simplest possible form. Thus the motion of the nodes is described in terms of the transverse displacement component w_j and the rotation θ_j, and the corresponding nodal force only includes the transverse force component P_j and the bending moment M_j. If beam elements are joined at an angle, as e.g. in Fig. 2.25, it is convenient to introduce the axial displacement component u_j and the axial force N_j into the formulation. This extension of the theory is straightforward, but it adds little to the theory of the individual flexible structural elements, and it is therefore not discussed further here. Axial force and displacement are discussed in connection with beam-columns in Chapter 3, but for a general discussion of the role in finite element analysis the reader is referred to specialized texts like Cook et al. (1989) or Thelanderson (1984).

2.7   Shear flexibility

The basic assumption of the Bernoulli beam theory, summarized in Box 2.3, is that the only deformation mechanism is curvature. This mechanism was identified with reference to constant bending moment, illustrated in Figs. 2.2 and 2.4. In this particular case the rotation of the cross-section is identical to the rotation of the tangent of the beam axis, and thus sections initially orthogonal to the beam axis will remain orthogonal to the beam axis in the deformed state. However, this is an approximation, in which the effect of shear deformation is neglected. Figure 2.36 shows the bending and the shear deformation modes.

The basic parameters are the rotation of the cross-section $\theta(x)$ and the rotation of the tangent of the beam axis $-dw(x)/dx$. In bending fibers parallel to the beam axis change length due to a change of the cross-section rotation $\theta(x)$ along the beam. The change of angle between neighboring cross-sections is $d\theta = \kappa\,dx$. Thus, the parameter κ associated with beam bending is the

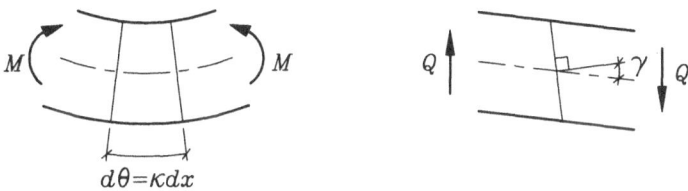

Figure 2.36: Deformation mechanisms in shear flexible
beams: a) bending, b) shear.

change in cross-section rotation per unit length. The shear mechanism accounts for the fact that a shear force Q in the beam will introduce shear
strains as explained in Section 1.6, and the 'average' shear strain γ appears
as an angle between the beam axis tangent and the cross-section normal as
shown in Fig. 2.36b. Thus, the kinematic relations for a beam theory including shear deformations consist of a definition of κ in terms of the angle θ and
a definition of the angle θ in terms of the transverse displacement w. The
shear mechanism in Fig. 2.36b shows that the rotation of a cross-section is
$-\mathrm{d}w/\mathrm{d}x$ due to rotation of the beam axis plus an additional rotation γ due
to shear straining. Thus, the appropriate kinematic relations are

$$\kappa = \frac{\mathrm{d}\theta}{\mathrm{d}x} \quad , \quad \theta = -\frac{\mathrm{d}w}{\mathrm{d}x} + \gamma \tag{2.79}$$

It is seen that the kinematic relations of Bernoulli beam theory are recovered
for $\gamma = 0$.

The beam bending mechanism, summarized in Box 2.1, remains unchanged,
when κ is defined as the gradient of the cross-section rotation angle, and thus
κ is proportional to the bending moment M. Similarly the shear strain γ is
proportional to the shear force Q. The constitutive relations, connecting the
generalized strains κ and γ with the internal forces M and Q, are

$$M = EI\kappa \quad , \quad Q = GA_0\gamma \tag{2.80}$$

EI is the bending stiffness, described in Section 2.1. GA_0 is the shear stiffness, consisting of the elastic shear modulus G and a representative cross-
section area A_0. As already mentioned γ is a representative average shear
strain, and A_0 is the corresponding area. Shear stresses are non-uniformly
distributed over the cross section, and this implies that the *shear area* A_0
is smaller than the full cross-section area A. For a rectangular cross-section

---

Box 2.10: Theory of beams with shear flexibility

The small displacement theory of beams with shear flexibility is based on statics (equilibrium),

$$\frac{dQ}{dx} = -p \quad , \quad \frac{dM}{dx} = Q$$

and the kinematic relations,

$$\kappa = \frac{d\theta}{dx} \quad , \quad \gamma = \frac{dw}{dx} + \theta$$

defining κ and the shear strain γ in terms of the transverse displacement w and the cross-section rotation angle θ.

In elastic beams the generalized strains κ and γ are proportional to the moment M and the shear force Q,

$$M = EI\kappa \quad , \quad Q = GA_0\gamma$$

where EI and GA_0 are the bending and shear stiffness, respectively.

$A_0 = \frac{5}{6}A$, and for an I-section the shear area A_0 is approximately equal to the area of the web. The basic relations of shear flexible beams are summarized in Box 2.10.

The boundary conditions of a beam are expressed in terms of the kinematic parameters w and θ or the internal forces Q and M. When shear flexibility is included, it is often most convenient to integrate the basic equilibrium and kinematic relations directly. This procedure is illustrated in the following two examples

Example 2.20

The effect of shear flexibility is illustrated by the cantilever beam loaded with a concentrated force P at the end as shown in Fig. 2.37. The cantilever is statically determinate, and the internal moment and shear force are

$$M(x) = (x - \ell)P \quad , \quad Q(x) = P$$

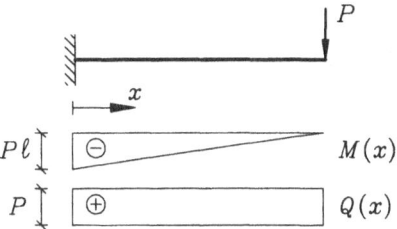

Figure 2.37: Cantilever beam with shear flexibility.

The bending relation for the cross-sections rotation $\theta(x)$,

$$\frac{\mathrm{d}\theta}{\mathrm{d}x} = \frac{M(x)}{EI} = \frac{P}{EI}(x - \ell)$$

is integrated to give

$$\theta = \frac{P}{EI}(\tfrac{1}{2}x^2 - x\ell) + C_1$$

The arbitrary constant C_1 is the value of θ at the support, and thus $C_1 = 0$.

The transverse displacement $w(x)$ is now determined by integrating the shear strain relation

$$\frac{\mathrm{d}w}{\mathrm{d}x} = -\theta + \gamma = \frac{P}{EI}(x\ell - \tfrac{1}{2}x^2) + \frac{P}{GA_0}$$

where the shear strain γ has been expressed in terms of the shear force P. Integration gives

$$w = \frac{P}{EI}(\tfrac{1}{2}x^2\ell - \tfrac{1}{6}x^3) + \frac{P}{GA_0}x + C_2$$

The arbitrary constant C_2 is the displacement w at the support, whereby $C_2 = 0$. Thus, the displacement of a shear flexible cantilever is

$$w = \frac{P\ell^3}{6EI}\left(\frac{x}{\ell}\right)^2\left[3 - \frac{x}{\ell}\right] + \frac{P\ell}{GA_0}\frac{x}{\ell}$$

It is seen, that the rotation of the cross-sections $\theta(x)$ is independent of the shear flexibility, while the displacement $w(x)$ consists of two additive contributions, a contribution from bending flexibility identical to that for Bernoulli beams, and a contribution from shear deformation. This additive form of the displacement remains valid for other statically determinate load cases.

The importance of including shear flexibility in the beam theory can be estimated by comparing bending and shear contributions, w_E and w_G, to the displacement at the tip of the beam. As seen from the solution

$$w_E = \frac{P\ell^3}{3EI} \quad , \quad w_G = \frac{P\ell}{GA_0}$$

Thus the relative importance of the shear contribution in the present case is determined by the ratio

$$\frac{w_G}{w_E} = \frac{3EI}{GA_0\ell^2}$$

The bending moment of inertia I is proportional to the cross-section area and the square of its height h, see Example 2.1. Thus, the ratio w_G/w_E is proportional to $(h/\ell)^2$, and the shear effect is seen to be most pronounced for high and short beams.

In the deformation and finite element methods distributed loads are included via their equivalent concentrated loads on the nodes. These concentrated loads correspond to the reactions at the ends of a rigidly supported beam, and it is therefore of particular interest to investigate the influence of shear flexibility on the reactions of a rigidly supported beam.

Example 2.21

Figure 2.38 shows a rigidly supported homogeneous beam with uniformly distributed load p. In this case it is convenient to use a coordinate x with origin at the center of the beam. The solution is obtained by sequential integration of the basic relations from Box 2.10.

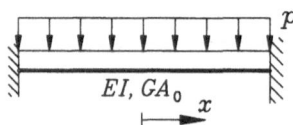

Figure 2.38: Rigidly supported beam with uniform load.

The shear force $Q(x)$ is obtained by integration of the transverse equilibrium equation,

$$\frac{dQ}{dx} = -p \quad \Rightarrow \quad Q(x) = -px$$

where the symmetry condition $Q(0) = 0$ has been used to eliminate an arbitrary constant. Integration of the moment relation gives

$$\frac{dM}{dx} = Q = -px \quad \Rightarrow \quad M(x) = -\tfrac{1}{2}p\,x^2 + M_0$$

where the arbitrary constant M_0 is the moment at the center of the beam. The bending relation then gives

$$\frac{d\theta}{dx} = \frac{M}{EI} = \frac{1}{EI}\left(-\tfrac{1}{2}p\,x^2 + M_0\right) \quad \Rightarrow \quad \theta(x) = \frac{1}{EI}\left(-\tfrac{1}{6}p\,x^3 + M_0 x\right)$$

where the symmetry condition $\theta(0) = 0$ has been used to eliminate an arbitrary constant. The moment M_0 is determined by the boundary condition $\theta(\tfrac{1}{2}\ell) = 0$, whereby $M_0 = \tfrac{1}{24}p\ell^2$. The center moment M_0, and thereby the full distribution of internal forces, are seen to be independent of shear flexibility.

Finally, the displacement field follows from integration of the shear relation, with $\gamma = Q/GA_0$.

$$\frac{dw}{dx} = -\theta + \gamma = -\frac{p}{24EI}\left(x\ell^2 - 4x^3\right) - \frac{p\,x}{GA_0}$$

from which

$$w(x) = -\frac{p}{24EI}\left(\tfrac{1}{2}x^2\ell^2 - x^4\right) - \frac{p\,x^2}{2GA_0} + w_0$$

The arbitrary constant w_0, representing the displacement at the center, is determined from the boundary condition $w(\tfrac{1}{2}\ell) = 0$,

$$w_0 - \frac{p\,\ell^4}{384EI} + \frac{p\,\ell^2}{8GA_0}$$

Substitution of this gives the displacement field

$$w(x) = \frac{p\,\ell^4}{384EI}\left[1 - 2\left(\frac{2x}{\ell}\right)^2 + \left(\frac{2x}{\ell}\right)^4\right] + \frac{p\,\ell^2}{8GA_0}\left[1 - \left(\frac{2x}{\ell}\right)^2\right]$$

Also in this case the displacement field is the sum of a bending and a shear deformation contribution. At the center the relative magnitude of the shear contribution is $w_G/w_E = 48EI/(GA_0\ell^2)$, and thus again the shear contribution to the deformation is determined by the non-dimensional parameter $EI/(GA_0\ell^2)$.

The examples illustrate, that in spite of the seemingly small change in the governing equations, integration of the equations of shear flexible beams is considerably more cumbersome than those of the simple Bernoulli bending beam theory. In practice it is often sufficient to know the beam element stiffness, and to be able to calculate the displacements at selected points. The most efficient procedure for this is the principles of virtual work and complementary virtual work. The principle of virtual work involves the actual static fields in combination with a virtual kinematic field, while the principle of complementary virtual work involves the actual kinematic fields in combination with a virtual static field. In their general form neither of these principles involve any assumption about the constitutive relations. However, here the main interest is the determination of displacements of elastic beams, and thus the actual and virtual fields satisfy the static as well as the kinematic relations. In that case the difference between the two principles is mainly a question of terminology. Here the principle of virtual work is derived for shear flexible beams, and the relevant formulas for displacements and rotations are obtained directly from this principle.

The principle of virtual work is based on the equilibrium equation

$$\frac{\mathrm{d}^2 M}{\mathrm{d}x^2} + p(x) = 0 \tag{2.81}$$

This equation is multiplied by a virtual displacement field $\delta w(x)$ and integrated over the beam length

$$\int_0^\ell \delta w(x) \Big(\frac{\mathrm{d}^2 M}{\mathrm{d}x^2} + p(x) \Big) \, \mathrm{d}x = 0 \tag{2.82}$$

Integration by parts of the first term gives

$$\Big[\delta w \, \frac{\mathrm{d}M}{\mathrm{d}x} \Big]_0^\ell + \int_0^\ell \Big(- \frac{\mathrm{d}(\delta w)}{\mathrm{d}x} \frac{\mathrm{d}M}{\mathrm{d}x} + \delta w \, p(x) \Big) \, \mathrm{d}x = 0 \tag{2.83}$$

Now, $\mathrm{d}M/\mathrm{d}x = Q$ and the cross-section rotation angle $\delta\theta$ is introduced according to (2.79b).

$$\Big[\delta w \, Q \Big]_0^\ell + \int_0^\ell \Big((\delta\theta - \delta\gamma) \frac{\mathrm{d}M}{\mathrm{d}x} + \delta w \, p(x) \Big) \, \mathrm{d}x = 0 \tag{2.84}$$

Integration by parts of the first term, and introduction of $\delta\kappa$ according to (2.79a), give

$$\Big[\delta w \, Q + \delta\theta \, M \Big]_0^\ell + \int_0^\ell \Big(- \delta\gamma \, Q - \delta\kappa \, M + \delta w \, p(x) \Big) \, \mathrm{d}x = 0 \tag{2.85}$$

After rearranging the terms and introducing the moments and forces with a
global sign convention as in Section 2.4, the following form of the principle
of virtual work is obtained

$$\sum_i \delta w_i \, P_i \; + \; \sum_j \delta\theta_j \, M_j \; + \; \int_0^\ell \delta w \, p \, \mathrm{d}x \; = \; \int_0^\ell (\delta\gamma \, Q + \delta\kappa \, M) \, \mathrm{d}x \qquad (2.86)$$

This is the equation of virtual work for a shear flexible beam. The corresponding principle of virtual work states that for an arbitrary virtual displacement
field, satisfying the kinematic relations (2.79), the virtual external work is
equal to the internal virtual work. The only difference from the virtual work
equation of Bernoulli beam bending theory is, that in the present case the
internal work includes a contribution from the work of the shear force, $\delta\gamma \, Q$,
as well as the previous work of the internal moment, $\delta\kappa \, M$.

In the particular case of elastic beams, the roles of virtual and actual fields
may be interchanged, combining the actual kinematic field with a virtual
static field. If the virtual static field corresponds to a concentrated virtual
force δP_i, the displacement w_i is given by the integral

$$\delta P_i \, w_i \; = \; \int_0^\ell (\delta Q \, \gamma + \delta M \, \kappa) \, \mathrm{d}x \; = \; \int_0^\ell \left(\frac{\delta Q \, Q}{G A_0} + \frac{\delta M \, M}{E I} \right) \mathrm{d}x \qquad (2.87)$$

and if the virtual static field corresponds to a concentrated virtual moment
δM_j the rotation θ_j is given by

$$\delta M_j \, \theta_j \; = \; \int_0^\ell (\delta Q \, \gamma + \delta M \, \kappa) \, \mathrm{d}x \; = \; \int_0^\ell \left(\frac{\delta Q \, Q}{G A_0} + \frac{\delta M \, M}{E I} \right) \mathrm{d}x \qquad (2.88)$$

It is seen from the integrals that the effect of shear flexibility appears in the
form of an extra term in the integral. For a statically determinate structure the internal force distribution is independent of the material properties,
and thus the effect of shear flexibility is simply an additive contribution to
displacements and cross-section rotations.

The principle of virtual work for shear flexible beams, and the complementary
work integrals for displacements and rotations are summarized in Box. 2.11.

Example 2.22

Consider the computation of the displacement at the tip of shear flexible
cantilever beam with a transverse load P at the tip by the principle of
complementary virtual work. The moment and shear force distributions
are shown in Fig. 2.37, and the tip displacement then follows directly

Box 2.11: Virtual work equation for a shear flexible beam

The internal moment $M(x)$ and shear force $Q(x)$ of a beam satisfy the static differential relations

$$\frac{dQ}{dx} = -p \quad , \quad \frac{dM}{dx} = Q$$

and discontinuity relations at the concentrated loads

$$Q(x_i^+) - Q(x_i^-) = -P_i \quad , \quad M(x_j^+) - M(x_j^-) = -M_j$$

Introduce a virtual displacement $\delta w(x)$ and define the virtual 'curvature' and cross-section rotation through the kinematic relations

$$\delta\kappa = \frac{d(\delta\theta)}{dx} \quad , \quad \delta\theta = -\frac{d(\delta w)}{dx} + \delta\gamma$$

where $\delta\gamma$ is the virtual shear strain.

For any choice of virtual displacements the external virtual work is equal to the internal virtual work,

$$\Sigma_i\,\delta w_i\,P_i + \Sigma_j\,\delta\theta_j\,M_j + \int_0^\ell \delta w\,p\,dx = \int_0^\ell (\delta\gamma\,Q + \delta\kappa\,M)\,dx$$

For an elastic shear flexible beam the deformation follows from the constitutive relations $M = EI\kappa$ and $Q = GA_0\gamma$. The displacement w_i can then be found from the virtual internal forces $\delta Q, \delta M$ corresponding to a virtual unit force $\delta P_i = 1$,

$$w_i = \int_0^\ell (\delta Q\,\gamma + \delta M\,\kappa)\,dx = \int_0^\ell \left(\frac{\delta Q\,Q}{GA_0} + \frac{\delta M\,M}{EI}\right)dx$$

The rotation θ_j of a cross-section can be found from the virtual internal forces $\delta Q, \delta M$ corresponding to a virtual unit moment $\delta M_j = 1$,

$$\theta_j = \int_0^\ell (\delta Q\,\gamma + \delta M\,\kappa)\,dx = \int_0^\ell \left(\frac{\delta Q\,Q}{GA_0} + \frac{\delta M\,M}{EI}\right)dx$$

from the complementary work of a transverse unit force, as given by (2.86),

$$w = \int_0^\ell \Big(\frac{\delta Q\, Q}{GA_0} + \frac{\delta M\, M}{EI} \Big)\, \mathrm{d}x = \frac{P\ell}{GA_0} + \frac{P\ell^3}{3EI}$$

where the result follows directly from the integration formulas in Appendix B. The same result was obtained from integration of the full solution in Example 2.20. Note, how directly the shear contribution to the displacement is included, when the static field is known.

The examples have illustrated that the shear deformation mechanism leads to additional deformations. Thus, the flexibility of a beam appears to consist of the sum of a bending contribution and a shear contribution. This property can be used to obtain a rather simple form of the stiffness matrix of a shear flexible beam. Figure 2.39a shows the four basic modes of deformation of a beam element. Each of these deformation modes may be determined by integration of the differential equations from Box 2.10, and the internal forces can then be used to determine the element stiffness matrix, following the direct method from Section 2.6. However, due to the form of the differential equations, this procedure is rather cumbersome. Also, use of the displacement shape functions and the principle of virtual work, requires computation of the correct shape functions in order to obtain the correct stiffness matrix.

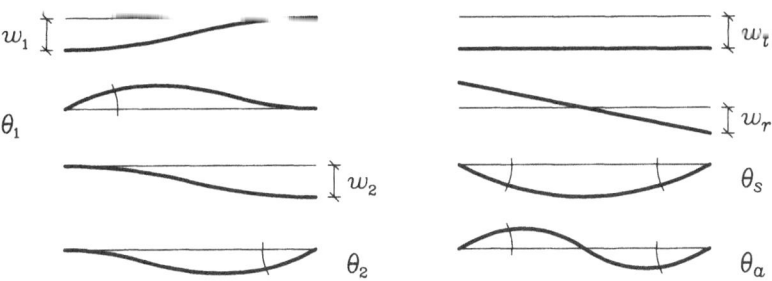

Figure 2.39: Displacement modes of beam element: a) unit displacements at nodes, b) rigid body and deformation modes.

An alternative method, based directly on the flexibility, is suggested by the displacement modes shown in Fig. 2.39b. The first two of these displacement modes represent a rigid body translation and rotation, respectively. The last two modes are deformation modes, corresponding to symmetric bending and anti-symmetric bending, respectively. It is seen, that only the last two modes lead to deformation of the beam element, and thus the stiffness of the beam can be evaluated from only these two displacement modes.

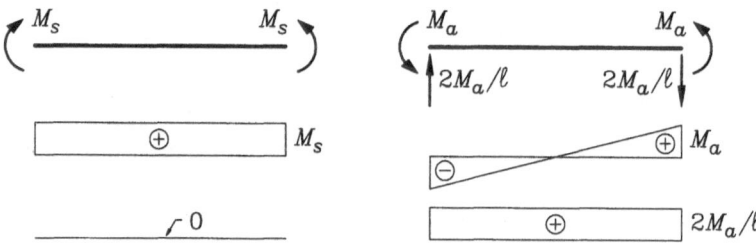

Figure 2.40: Deformation modes of beam element: a) symmetric bending, b) anti-symmetric bending.

Figure 2.40 shows the nodal forces and the corresponding internal shear force and moment for symmetric and anti-symmetric bending. Symmetric bending gives a constant internal moment, and therefore no shear force. In anti-symmetric bending moment equilibrium of the beam leads to equal and opposite transverse nodal forces, and thereby a constant shear force of magnitude $Q = 2M_a/\ell$. The rotation of the nodes can now be determined by application of a symmetric virtual moment pairs $\delta M_s = 1$ and an anti-symmetric virtual moment pair $\delta M_a = 1$, respectively. The external virtual work then gives equal contributions from each end, and the rotation of the nodes are found from the complementary virtual work integrals

$$2\,\theta_s \;=\; \int_0^\ell \frac{\delta M_s(x)\, M_s(x)}{EI}\;\mathrm{d}x \;=\; \frac{\ell\,M_s}{EI} \tag{2.89}$$

and

$$2\,\theta_a \;=\; \int_0^\ell \Big(\frac{\delta Q_s(x)\, Q_s(x)}{GA_0} + \frac{\delta M_s(x)\, M_s(x)}{EI} \Big) \mathrm{d}x \;=\; \frac{4\ell\,M_a}{GA_0\ell^2} + \frac{\ell\,M_a}{3EI} \tag{2.90}$$

where the integrals follow from the product rules in Appendix B. Thus the bending flexibility of the beam element is contained in the two relations

$$\theta_s \;=\; \frac{\ell\,M_s}{2EI} \quad , \quad \theta_a \;=\; (1+\Phi)\frac{\ell\,M_a}{6EI} \tag{2.91}$$

where the effect of the shear deformation is contained in the parameter

$$\Phi = \frac{12EI}{GA_0\ell^2} \tag{2.92}$$

In Bernoulli beam theory $\Phi = 0$.

The nodal force vector $f^T = [\,P_1, M_1, P_2, M_2\,]$ is now expressed in terms of the moments M_s and M_a by direct observations from Fig. 2.40.

$$\begin{bmatrix} P_1 \\ M_1 \\ P_2 \\ M_2 \end{bmatrix} = \begin{bmatrix} -2M_a/\ell \\ -M_s + M_a \\ 2M_a/\ell \\ M_s + M_a \end{bmatrix} = \frac{1}{\ell} \begin{bmatrix} 0 & -2 \\ -\ell & \ell \\ 0 & 2 \\ \ell & \ell \end{bmatrix} \begin{bmatrix} M_s \\ M_a \end{bmatrix} \tag{2.93}$$

Substitution of the moments $[\,M_s, M_a\,]$ from (2.91) then gives the nodal force vector in terms of the angles $[\,\theta_s, \theta_a\,]$ from the symmetric and anti-symmetric deformation modes.

$$\begin{bmatrix} P_1 \\ M_1 \\ P_2 \\ M_2 \end{bmatrix} = \frac{2EI}{(1+\Phi)\ell^2} \begin{bmatrix} 0 & -6 \\ -(1+\Phi)\ell & 3\ell \\ 0 & 6 \\ (1+\Phi)\ell & 3\ell \end{bmatrix} \begin{bmatrix} \theta_s \\ \theta_a \end{bmatrix} \tag{2.94}$$

All that now remains is to express the angles $[\,\theta_s, \theta_a\,]$ in terms of the nodal displacement vector $u^T = [\,w_1, \theta_1, w_2, \theta_2\,]$. The angles $[\,\theta_s, \theta_a\,]$ are determined as the symmetric and anti-symmetric part of the node rotations, after subtraction of the rigid body rotation $\theta_r = -(w_2 - w_1)/\ell$. Thus the angles $[\,\theta_s, \theta_a\,]$ are

$$\theta_s = \tfrac{1}{2}(\theta_2 - \theta_1) \tag{2.95}$$

$$\theta_a = \tfrac{1}{2}(\theta_2 + \theta_1) + (w_2 - w_1)/\ell \tag{2.96}$$

Substitution of these expressions into (2.94) gives the stiffness matrix relation

$$\begin{bmatrix} P_1 \\ M_1 \\ P_2 \\ M_2 \end{bmatrix} = \frac{EI}{(1+\Phi)\ell^3} \begin{bmatrix} 12 & -6\ell & -12 & -6\ell \\ -6\ell & (4+\Phi)\ell^2 & 6\ell & (2-\Phi)\ell^2 \\ -12 & 6\ell & 12 & 6\ell \\ -6\ell & (2-\Phi)\ell^2 & 6\ell & (4+\Phi)\ell^2 \end{bmatrix} \begin{bmatrix} w_1 \\ \theta_1 \\ w_2 \\ \theta_2 \end{bmatrix} \tag{2.97}$$

summarized in Box 2.12. It is seen that the stiffness matrix of the Bernoulli beam element, given in Box. 2.9, is recovered for $\Phi = 0$.

Box 2.12: Stiffness matrix of shear flexible beam element

The stiffness matrix K of a beam element establishes a relation of the form $\mathsf{f} = \mathsf{K}\,\mathsf{u}$ between node displacement components u and corresponding nodal force components f,

$$\mathsf{u}^T \;=\; [\,w_1, \theta_1, w_2, \theta_2\,] \qquad , \qquad \mathsf{f}^T \;=\; [\,P_1, M_1, P_2, M_2\,]$$

For a shear flexible beam element with bending stiffness EI and shear stiffness GA_0 - see Box 2.10 - the stiffness matrix is

$$\mathsf{K} \;=\; \frac{EI}{(1+\Phi)\ell^3}
\begin{bmatrix}
12 & -6\ell & -12 & -6\ell \\
-6\ell & (4+\Phi)\ell^2 & 6\ell & (2-\Phi)\ell^2 \\
-12 & 6\ell & 12 & 6\ell \\
-6\ell & (2-\Phi)\ell^2 & 6\ell & (4+\Phi)\ell^2
\end{bmatrix}$$

where the effect of the shear flexibility is described by the non-dimensional parameter

$$\Phi \;=\; \frac{12EI}{GA_0\ell^2}$$

The present method of deriving the stiffness matrix relies on separation of the displacement fields into rigid body modes and deformation modes. The deformation modes can then be analyzed in terms of flexibility, and the results combined into the stiffness matrix format. This approach can also be used to analyze more general problems, such as curved beams or beams with variable bending and shear stiffness, Krenk (1994).

2.8 Beams on elastic foundation

Beams with distributed elastic lateral support are known as beams on elastic foundation. This terminology arises from early applications of this theory to foundation problems, but there are several other applications within beam theory itself. Thus, the elastic lateral support may describe the wall of an axisymmetric cylindrical shell or the resistance to distortion of a box girder. Numerous specific solutions have been discussed by Hetenyi (1946), and extended models have been reviewed by Keer (1964). The present section is limited to a presentation of the basic differential equations and some important special solutions.

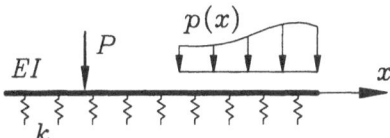

Figure 2.41: Beam on simple spring foundation.

The basic problem is shown in Fig. 2.41. A beam with bending stiffness EI is supported laterally on continuously distributed elastic springs with stiffness k per unit beam length. These springs give a transverse load $p(x) = -kw(x)$ in addition to the external load $p(x)$. Thus, the beam bending equation (2.21) takes the form

$$\frac{\mathrm{d}^2}{\mathrm{d}x^2}\left(EI\,\frac{\mathrm{d}^2w}{\mathrm{d}x^2}\right) + k\,w(x) = p(x) \tag{2.98}$$

For constant bending and spring stiffness the differential equation can be written as

$$\frac{\mathrm{d}^4w}{\mathrm{d}x^4} + 4\lambda^4\,w(x) = \frac{p(x)}{EI} \tag{2.99}$$

with the length-scale parameter

$$\lambda = \sqrt[4]{\frac{k}{4EI}} \tag{2.100}$$

Solutions to the homogeneous equation of the form $w(x) \propto \exp(\alpha x)$ are obtained from the characteristic equation

$$\alpha^4 + 4\lambda^4 = 0 \tag{2.101}$$

with the solutions

$$\alpha = \pm\lambda(1 \pm i) \tag{2.102}$$

Here i is the imaginary unit, and α a complex number. The complete solution to the homogeneous equation is conveniently written in terms of exponentials and trigonometric functions as

$$
\begin{aligned}
w(x) = \; & e^{-\lambda x}\left[\, C_1 \cos(\lambda x) + C_2 \sin(\lambda x)\,\right] \\
& + e^{\lambda x}\left[\, C_3 \cos(\lambda x) + C_4 \sin(\lambda x)\,\right]
\end{aligned}
\tag{2.103}
$$

where $C_1, \cdots, C_4$ are arbitrary constants to be determined by the boundary conditions. The solution depends on the coordinate x only through the non-dimensional product λx, and thus $1/\lambda$ serves as a characteristic length.

Example 2.23

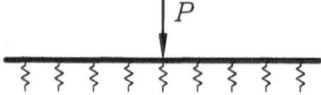

Figure 2.42: Infinite beam with concentrated force.

Consider a beam of infinite length with a concentrated force at $x = 0$, Fig. 2.42. Due to symmetry only the half $x \geq 0$ needs to be considered. In the solution (2.103) the two last terms are unbounded for $x \to \infty$, and thus the present solution is of the form

$$w(x) = e^{-\lambda x}\left[\, C_1 \cos(\lambda x) + C_2 \sin(\lambda x)\,\right] \;,\quad x \geq 0$$

The solution must satisfy the condition $dw/dx = 0$ for $x = 0$, whereby $C_1 = C_2$, and

$$w(x) = C_1 e^{-\lambda x}\left[\, \cos(\lambda x) + \sin(\lambda x)\,\right] \;,\quad x \geq 0$$

The constant C_1 follows from equilibrium with the applied force P. Half of the force must be carried by the springs of the semi-infinite beam $x > 0$,

$$\tfrac{1}{2}P = \int_0^\infty k\, w(x)\, dx = k\, C_1 \frac{1}{\lambda}\Big[-e^{-\lambda x} \cos(\lambda x)\Big]_0^\infty = k\, C_1 \frac{1}{\lambda}$$

This determines the arbitrary constant $C_1 = P\lambda/2k$, giving the displacement

$$w(x) = \frac{P\lambda}{2k} e^{-\lambda x} [\cos(\lambda x) + \sin(\lambda x)] \quad , \quad x \geq 0$$

The rest of the solution for $x \geq 0$ now follows from differentiation

$$\theta(x) = -\frac{dw}{dx} = \frac{P\lambda^2}{k} e^{-\lambda x} \sin(\lambda x) \quad , \quad x \geq 0$$

$$M(x) = EI\frac{d\theta}{dx} = \frac{P}{4\lambda} e^{-\lambda x} [\cos(\lambda x) - \sin(\lambda x)] \quad , \quad x \geq 0$$

$$Q(x) = \frac{dM}{dx} = -\frac{P}{2} e^{-\lambda x} \cos(\lambda x) \quad , \quad x > 0$$

The solution is illustrated in Fig. 2.43

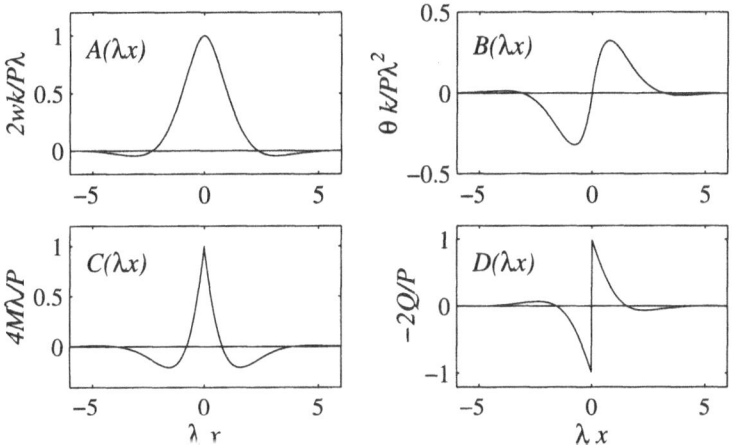

Figure 2.43: Solution for concentrated force on infinite beam.

The functions defining $w(x)$, $\theta(x)$, $M(x)$ and $Q(x)$ are conveniently used to express solutions for infinite and semi-infinite beams on elastic foundation. They are introduced as $A(x), B(x), C(x), D(x)$ in Box 2.13. In terms of these functions the solution is

$$w(x) = \frac{P\lambda}{2k} A(\lambda x) \qquad \theta(x) = \frac{P\lambda^2}{k} B(\lambda x)$$

$$M(x) = \frac{P}{4\lambda} C(\lambda x) \qquad Q(x) = -\frac{P}{2} D(\lambda x)$$

Box 2.13: Basis functions for beam on elastic springs

The homogeneous equation for a beam on elastic springs

$$\frac{\mathrm{d}^4 w}{\mathrm{d}x^4} + 4\lambda^4\, w(x) = 0$$

with parameter $\lambda^4 = k/4EI$ is conveniently solved in terms of four functions, given for $y \geq 0$ as

$$A(y) = \mathrm{e}^{-y}[\,\cos(y) + \sin(y)\,] \quad , \quad B(y) = \mathrm{e}^{-y}\sin(y)$$

$$C(y) = \mathrm{e}^{-y}[\,\cos(y) - \sin(y)\,] \quad , \quad D(y) = \mathrm{e}^{-y}\cos(y)$$

For $y < 0$ the functions are defined by

$$A(y) = A(-y) \quad , \quad B(y) = -B(-y)$$

$$C(y) = C(-y) \quad , \quad D(y) = -D(-y)$$

These functions satisfy the cyclic differentiation relations

$$\frac{\mathrm{d}A(y)}{\mathrm{d}y} = -2B(y) \quad , \quad \frac{\mathrm{d}B(y)}{\mathrm{d}y} = C(y)$$

$$\frac{\mathrm{d}C(y)}{\mathrm{d}y} = -2D(y) \quad , \quad \frac{\mathrm{d}D(y)}{\mathrm{d}y} = -A(y)$$

The functions $A(y), B(y), C(y), D(y)$ are shown in Fig. 2.43.

Example 2.24

The problem of an infinite beam with spring supports acted on by a concentrated moment follows from the concentrated force solution by considering two forces of magnitude $-P$ and P, acting at $x = \frac{1}{2}a$ and $x = -\frac{1}{2}a$, respectively. At $x > \frac{1}{2}a$ the displacement follows from the solution of the previous example by superposition,

$$w(x) = \frac{P\lambda}{2k}\left[A(\lambda(x + \tfrac{1}{2}a)) - A(\lambda(x - \tfrac{1}{2}a)) \right]$$

The solution for the concentrated moment is obtained by considering the limit $a \to 0$, while keeping $M_0 = Pa$ constant.

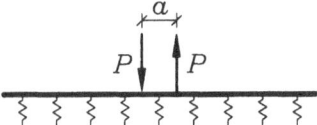

Figure 2.44: Infinite beam with concentrated moment $M_0 = Pa$.

$$w(x) \;=\; \frac{M_0\lambda}{2k}\frac{\mathrm{d}}{\mathrm{d}x}A(\lambda x) \;=\; -\,\frac{M_0\lambda^2}{k}B(\lambda x)$$

where the last equality follows from the differentiation rule in Box 2.13. The rest of the concentrated moment solution follows from the differentiation rules in the form

$$\theta(x) \;=\; \frac{M_0\lambda^3}{k}C(\lambda x) \;,\quad M(x) \;=\; -\tfrac{1}{2}M_0D(\lambda x) \;,\quad Q(x) \;=\; \tfrac{1}{2}\lambda M_0 A(\lambda x)$$

Also this solution is illustrated by the graphs in Fig. 2.43.

The functions $A(y), B(y), C(y), D(y)$ defined in Box 2.13 form a convenient basis for solutions to semi-infinite beams on elastic springs. Figure 2.45 shows a semi-infinite beam on elastic springs, loaded at the end $x = 0$ with either imposed deformations $[\,w_0, \theta_0\,]$ or loads $[\,P_0, M_0\,]$. The solution for imposed displacements at the end of a semi-infinite beam can be extracted from the solutions for concentrated force and moment on an infinite beam, obtained in Examples 2.23 and 2.24. On an infinite beam the concentrated force leads to a displacement w_0 without rotation at the point of application, and a concentrated moment leads to a rotation θ_0 without displacement of the point of application. These two modes of deformation are decribed by the functions $A(\lambda x)$ and $B(\lambda x)$, respectively. The displacement of a semi-infinite beam with imposed end motion is then seen to be

$$w(x) \;=\; w_0\,A(\lambda x) \;-\; \theta_0\,\lambda^{-1}\,B(\lambda x) \tag{2.104}$$

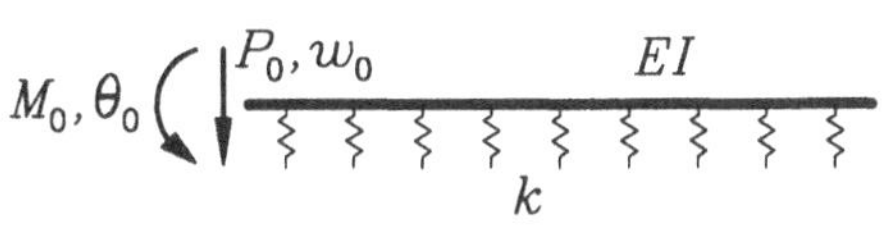

Figure 2.45: Semi-infinite beam with imposed end loads.

The rotation $\theta(x)$ and the internal forces $M(x), Q(x)$ now follow by differentiation, using the relations in Box 2.13. The stiffness matrix of the semi-infinite beam then follows from evaluation of the internal forces at the end $x = 1$ as in (2.65) for the beam without spring support. The results are summarized in Box 2.14.

Box 2.14: Semi-infinite beams on elastic springs

The displacement of a semi-infinite beam on elastic springs with imposed end motion $[\,w_0, \theta_0\,]$, shown in Fig. 2.45, is

$$w(x) \;=\; w_0\, A(\lambda x) \;-\; \theta_0\, \lambda^{-1}\, B(\lambda x)$$

The internal forces follow from differentiation as

$$\begin{bmatrix} M(x) \\ Q(x) \end{bmatrix} \;=\; \sqrt{k\,EI} \begin{bmatrix} C(\lambda x) & -\lambda^{-1}D(\lambda x) \\ -2\lambda\, D(\lambda x) & A(\lambda x) \end{bmatrix} \begin{bmatrix} w_0 \\ \theta_0 \end{bmatrix}$$

The stiffness matrix of the semi-infinite beam then follows from $[\,P_0, M_0\,] = -[\,Q(0), M(0)\,]$ as

$$\begin{bmatrix} P_0 \\ M_0 \end{bmatrix} \;=\; \sqrt{k\,EI} \begin{bmatrix} 2\lambda & -1 \\ -1 & \lambda^{-1} \end{bmatrix} \begin{bmatrix} w_0 \\ \theta_0 \end{bmatrix}$$

For applied end forces $[\,P_0, M_0\,]$ the end displacements follow from the inverse relation

$$\begin{bmatrix} w_0 \\ \theta_0 \end{bmatrix} \;=\; \frac{1}{\sqrt{k\,EI}} \begin{bmatrix} \lambda^{-1} & 1 \\ 1 & 2\lambda \end{bmatrix} \begin{bmatrix} P_0 \\ M_0 \end{bmatrix}$$

and the corresponding displacement is

$$w(x) \;=\; \frac{1}{\sqrt{k\,EI}} \Big[\, P_0\, \lambda^{-1}\, D(\lambda x) \;+\; M_0\, C(\lambda x) \,\Big]$$

Internal forces follow by the differentiation rules in Box 2.13.

$$\begin{bmatrix} M(x) \\ Q(x) \end{bmatrix} \;=\; \begin{bmatrix} -\lambda^{-1}B(\lambda x) & -A(\lambda x) \\ -C(\lambda x) & 2\lambda\, B(\lambda x) \end{bmatrix} \begin{bmatrix} P_0 \\ M_0 \end{bmatrix}$$

Cylindrical shells

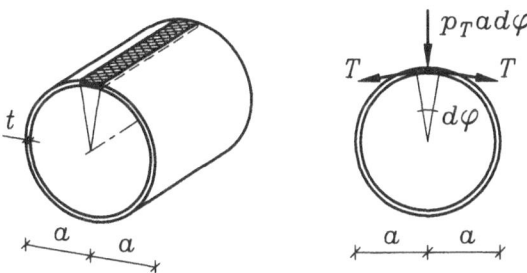

Figure 2.46: Elastic spring analogy for cylindrical shell.

An interesting application of the beam with distributed elastic spring support is axisymmetric problems for thin cylindrical shells. The mechanism is illustrated in Fig. 2.46. The cylindrical shell has a radius a and thickness t. A uniform inward motion w of the shell leads to a circumferential strain of magnitude $\varepsilon = -w/a$ in the shell, where the minus indicates compression. By Hooke's law this gives a hoop stress of magnitude $\sigma = E\varepsilon$, and the corresponding hoop force per unit length of the cylinder then is

$$T = \sigma t = -\frac{E t}{a} w \tag{2.105}$$

Equilibrium of an angular section, shown in Fig. 2.46, demonstrates the equivalence between a positive hoop force T and an inward radial force per unit shell area of magnitude

$$p_T = \frac{T}{a} = -\frac{E t}{a^2} w \tag{2.106}$$

This is a spring relation of the form $p_T = -kw$ with spring constant

$$k = \frac{E t}{a^2} \tag{2.107}$$

In axisymmetric deformation of a cylindrical shell the cross-sections are constrained by symmetry to move radially. This leads to an increased bending stiffness, corresponding to replacing the uniaxial elastic modulus E with $E/(1 - \nu^2)$, where ν is the Poisson ratio. The bending stiffness of a unit width of the shell then is, see Example 2.1,

$$\frac{E I}{1 - \nu^2} = \frac{E}{1 - \nu^2} \frac{t^3}{12} \tag{2.108}$$

The length-scale parameter λ for axisymmetric deformation of a cylindrical shell now follows from (2.100) as

$$\lambda \; = \; \sqrt[4]{\frac{k}{4}\frac{1-\nu^2}{EI}} \; = \; \sqrt[4]{3(1-\nu^2)}\,\frac{1}{\sqrt{at}} \tag{2.109}$$

For steel $\nu = 0.3$, giving the representative value $\lambda = 1.285/\sqrt{at}$. Thus, the characteristic length $\lambda^{-1} \simeq 0.8\sqrt{at}$ of the axisymmetric deformation pattern a thin cylindrical shell is considerably shorter than the radius a of the shell. This means that in long and thin cylindrical shells deformations induced at the ends of the shell will have the character of boundary effects as illustrated by the next example.

A more general discussion of the deformation of shells can be found e.g. in Calladine (1983) and Niordson (1985). In particular it is noteworthy that cylindrical shells may contain deformation modes with short wavelength, like the present axisymmetric modes, and deformation modes with long wavelength, like deformation into oval shape. The short wavelength deformation modes have a typical length $\lambda^{-1} \propto a(t/a)^{1/2}$, while the long wavelength modes have a typical length $\lambda^{-1} \propto a(a/t)^{1/2}$. Thus, for thin shells the axisymmetric deformation modes are very local, while the deformation of the cross-section into oval shape spreads over a long distance. The deformation of the shape of thin-walled box girders is of a similar long wavelength type.

Example 2.25

Figure 2.47 shows a cylindrical fluid filled tank of height h and wall thickness t. When the tank is filled with a fluid of specific weight γ, the fluid pressure on the wall is $\gamma(h-x)$. Thus, the deformation of the wall is governed by the differential equation

$$\frac{EI}{1-\nu^2}\frac{\mathrm{d}^4 w}{\mathrm{d}x^4} \; + \; \frac{Et}{a^2}\,w(x) \; = \; -\gamma(h-x)$$

The complete solution is obtained in the form $w(x) = w_\gamma(x) + w_b(x)$, where $w_\gamma(x)$ is the particular solution corresponding to the fluid pressure, and $w_b(x)$ represents the part of the solutions due to the imposed boundary conditions. A particular solution to the non-homogeneous equation is

$$w_\gamma(x) \; = \; -\frac{\gamma a^2}{Et}\,(h-x)$$

This is the linear displacement of the wall shown in Fig. 2.46b. The

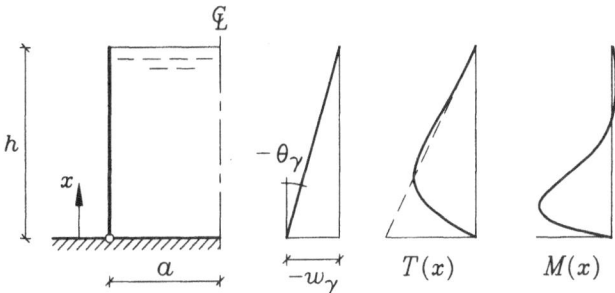

Figure 2.47: Fluid filled tank with simply supported wall.
$T(x)$ and $M(x)$ for $\lambda h = 5$.

displacement and rotation at ground level from this solution are

$$w_\gamma = -\frac{\gamma a^2}{E\,t}\,h \quad , \quad \theta_\gamma = -\frac{\gamma a^2}{E\,t}$$

In the present problem the tank wall has a simple support at ground level, and thus the boundary condition imposes a concentrated force P_0 of sufficient magnitude to eliminate the horizontal movement.

When it is assumed that the height is sufficient to eliminate influence from the top of the tank, the boundary solution $w_b(x)$ follows from the edge force solution of a semi-infinite beam in Box 2.14 as

$$w_b(x) = \frac{2\lambda}{k}\,D(\lambda x)\,P_0 = 2\lambda\,\frac{a^2}{E\,t}\,D(\lambda x)\,P_0$$

The magnitude of the support force P_0 follows from the condition $w(0) = w_\gamma(0) + w_b(0) = 0$, whereby

$$P_0 = -\frac{1}{2\lambda}\,\frac{E\,t}{a^2}\,w_\gamma(0) = \frac{\gamma\,h}{2\lambda}$$

With this edge force, the full wall displacement is

$$w(x) = \frac{\gamma\,a^2}{E\,t}\Big[-(h-x) + h\,D(\lambda x)\Big]$$

The corresponding hoop force $T(x)$ follows from the relation (2.106),

$$T(x) = -\frac{E\,t}{a}\,w(x) = \gamma\,a\Big[(h-x) - h\,D(\lambda x)\Big]$$

The distribution of the hoop force $T(x)$ is shown in Fig. 2.47c.

The internal moment $M(x)$ follows by differentiating the displacement $w(x)$ twice, or directly from the last formula in Box 2.14.

$$M(x) = -\frac{1}{\lambda} B(\lambda x) P_0 = -\frac{\gamma h}{2\lambda^2} B(\lambda x)$$

The moment distribution is shown in Fig. 2.47d. The numerically largest value M_{max} occurs at $dB(\lambda x)/dx = 0$, i.e. $\lambda x = \frac{1}{4}\pi$.

$$M_{max} = \frac{\gamma h}{2\lambda^2} B(\tfrac{1}{4}\pi) = \frac{\gamma h a t}{\sqrt{12(1-\nu^2)}} B(\tfrac{1}{4}\pi) = 0.093 \frac{\gamma h a t}{\sqrt{1-\nu^2}}$$

Beams of finite length

Solutions to problems involving beams of finite length can be obtained by superposition of a particular integral to the equilibrium equation (2.98) and four special solutions to account for the proper boundary conditions at the ends of the beam. It is important that these special solutions are selected with a view to the boundary conditions to be imposed. In the following the theory is developed in connection with the stiffness matrix of a beam with elastic spring support, i.e. for kinematic boundary conditions. By a slight modification an equally systematic procedure can be developed for static boundary conditions.

In the case of beams of infinite length the solution was greatly simplified by the condition that the solution is bounded at infinity. This reduced the number of arbitrary constants in the solution from four to two, and led to the basis functions $A(y), B(y), C(y), D(y)$ defined in Box 2.13. The convenience of these functions is due to two sets of properties: the cyclic differentiation rules, and the fact that one of the functions vanishes at $x = 0$. Furthermore, for 'long' beams with $\lambda \ell \gg 1$, the solution will have boundary layer character, as illustrated in the cylinder tank example. This implies that two terms are associated with each boundary, and the coupling between the boundaries is very weak. This may lead to ill-conditioned equations for the arbitrary constants for $\lambda \ell \gg 1$.

With these properties it is easy to uncouple the boundary conditions at $x = 0$. For beams of finite length the problem is complicated by the fact that boundary conditions must be imposed at both ends of the beam. However, it turns out that an explicit procedure can be developed by using a set of basis functions, suitable for beams of finite length.

In the case of beams of finite length ℓ a suitable set of basis functions is introduced by the following argument. The origin of the coordinate system at the left end of the beam, and an auxiliary coordinate defined by

$$x' = \ell - x \qquad (2.110)$$

The idea now is to replace the exponential function $\exp(-\lambda x)$ in the basis functions with one of the hyperbolic functions

$$\cosh(\lambda x') = \frac{e^{\lambda x'} + e^{-\lambda x'}}{2} \quad , \quad \sinh(\lambda x') = \frac{e^{\lambda x'} - e^{-\lambda x'}}{2} \qquad (2.111)$$

For 'long' beams the hyperbolic functions appear as a decreasing exponential around $x = 0$,

$$\cosh(\lambda x') \simeq \sinh(\lambda x') \simeq \tfrac{1}{2}e^{\lambda x'} = \tfrac{1}{2}e^{\lambda \ell}e^{-\lambda x} \quad , \quad x \ll \ell \qquad (2.112)$$

In terms of the variables $y = \lambda x$ and $y' = \lambda x'$ the basis functions defined in this way are

$$A_*(y) = \sinh(y')\cos(y) + \cosh(y')\sin(y) \quad , \quad B_*(y) = \sinh(y')\sin(y)$$

$$C_*(y) = \sinh(y')\cos(y) - \cosh(y')\sin(y) \quad , \quad D_*(y) = \cosh(y')\cos(y)$$
$$(2.113)$$

These basis functions satisfy the same differential relations as their counterparts in Box 2.13. Furthermore it is noted that the function $B_*(y)$ vanishes at *both* ends of the beam, $B_*(0) = B_*(\lambda\ell) = 0$.

The solution for a beam element of finite length ℓ with end forces and moments is decomposed into a symmetric part $w_s(x)$ and an anti-symmetric

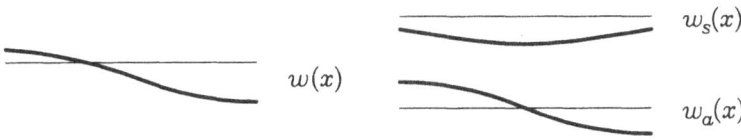

Figure 2.48: Full solution of beam element as sum of a symmetric solution $w_s(x)$ and an anti-symmetric solution $w_a(x)$.

part $w_a(x)$ as shown in Fig. 2.48. The symmetric part of the solution can be
written down directly in terms of the basis functions (2.113) as

$$w_s(x) \;=\; \frac{A_*(\lambda x) + A_*(\lambda x')}{A_*(\lambda\ell) + A_*(0)}\, w_s(\ell) \;-\; \frac{B_*(\lambda x) + B_*(\lambda x')}{C_*(\lambda\ell) - C_*(0)}\, \frac{1}{\lambda}\, \theta_s(\ell) \quad (2.114)$$

Clearly (2.114) is a solution, and the special form with equal contributions for
x and x' makes the solution symmetric. The fact that $B_*(0) = B_*(\lambda\ell) = 0$
establishes the value $w_s(\ell)$, and the differential rules for $A_*(y)$ and $B_*(y)$
implies that $dw_s(\ell)/dx = -\theta_s(\ell)$. The anti-symmetric solution is obtained
simply by changing the sign between the terms with x and x'.

$$w_a(x) \;=\; \frac{A_*(\lambda x) - A_*(\lambda x')}{A_*(\lambda\ell) - A_*(0)}\, w_a(\ell) \;-\; \frac{B_*(\lambda x) - B_*(\lambda x')}{C_*(\lambda\ell) + C_*(0)}\, \frac{1}{\lambda}\, \theta_a(\ell) \quad (2.115)$$

It is noted that neither of the four denominators

$$A_*(\lambda\ell) \pm A_*(0) \;=\; C_*(\lambda\ell) \mp C_*(0) \;=\; \sin(\lambda\ell) \pm \sinh(\lambda\ell) \quad (2.116)$$

vanish for any finite beam length.

The full solution is obtained as the sum of the symmetric and the anti-
symmetric parts, and this in turn implies that the end displacements satisfy
the relations

$$w_s(\ell) \;=\; \tfrac{1}{2}(w_2 + w_1) \;\;,\;\; \theta_s(\ell) \;=\; \tfrac{1}{2}(\theta_2 - \theta_1)$$
$$$$
$$w_a(\ell) \;=\; \tfrac{1}{2}(w_2 - w_1) \;\;,\;\; \theta_a(\ell) \;=\; \tfrac{1}{2}(\theta_2 + \theta_1). \quad (2.117)$$

Substitution of these expressions then gives the full solution

$$\begin{aligned}
w(x) \;=\;& \frac{A_*(\lambda x) + A_*(\lambda x')}{A_*(\lambda\ell) + A_*(0)}\, \frac{w_2 + w_1}{2} \;-\; \frac{B_*(\lambda x) + B_*(\lambda x')}{C_*(\lambda\ell) - C_*(0)}\, \frac{\theta_2 - \theta_1}{2\lambda} \\
+\;& \frac{A_*(\lambda x) - A_*(\lambda x')}{A_*(\lambda\ell) - A_*(0)}\, \frac{w_2 - w_1}{2} \;-\; \frac{B_*(\lambda x) - B_*(\lambda x')}{C_*(\lambda\ell) + C_*(0)}\, \frac{\theta_2 + \theta_1}{2\lambda}
\end{aligned}$$
$$(2.118)$$

By regrouping the terms it is seen that this is a solution expressed in terms
of the displacement vector $u^T = [\,w_1, \theta_1, w_2, \theta_2\,]$ and suitable shape functions,
each representing the transverse displacement corresponding to a unit nodal
displacement. This generalizes the results of Section 2.6 for the deformation
of a free beam.

The rotation angle $\theta(x)$, the internal moment $M(x)$ and the shear force $Q(x)$
can now be obtained from the displacement $w(x)$ by differentiation, using

---

Box 2.15: Stiffness matrix for beam on elastic springs

The stiffness matrix K of a beam of length ℓ and bending stiffness EI, supported on distributed springs with stiffness parameter k, is

$$\mathsf{K} = \frac{EI}{\ell^3} \begin{bmatrix} 12\psi_1 & -6\psi_3\ell & -12\psi_2 & -6\psi_4\ell \\ -6\psi_3\ell & 4\psi_5\ell^2 & 6\psi_4\ell & 2\psi_6\ell^2 \\ -12\psi_2 & 6\psi_4\ell & 12\psi_1 & 6\psi_3\ell \\ -6\psi_4\ell & 2\psi_6\ell^2 & 6\psi_3\ell & 4\psi_5\ell^2 \end{bmatrix}$$

where the coefficients $\psi_1, \cdots, \psi_6$ are as

$$\psi_1 = \tfrac{1}{3}(\lambda\ell)^2 \,\Psi\,\big[\sinh(\lambda\ell)\cosh(\lambda\ell) + \sin(\lambda\ell)\cos(\lambda\ell)\big]$$

$$\psi_2 = \tfrac{1}{3}(\lambda\ell)^2 \,\Psi\,\big[\sin(\lambda\ell)\cosh(\lambda\ell) + \sinh(\lambda\ell)\cos(\lambda\ell)\big]$$

$$\psi_3 = \tfrac{1}{3}(\lambda\ell)\,\Psi\,\big[\sinh^2(\lambda\ell) + \sin^2(\lambda\ell)\big]$$

$$\psi_4 = \tfrac{2}{3}(\lambda\ell)\,\Psi\,\sin(\lambda\ell)\,\sinh(\lambda\ell)$$

$$\psi_5 = \tfrac{1}{2}\,\Psi\,\big[\sinh(\lambda\ell)\cosh(\lambda\ell) - \sin(\lambda\ell)\cos(\lambda\ell)\big]$$

$$\psi_6 = \Psi\,\big[\sin(\lambda\ell)\cosh(\lambda\ell) - \sinh(\lambda\ell)\cos(\lambda\ell)\big]$$

with the common factor

$$\Psi = \frac{\lambda\ell}{\sinh^2(\lambda\ell) - \sin^2(\lambda\ell)}$$

---

the differentiation rules of Box 2.13. The process is straightforward and will not be given here. The nodal force vector $\mathsf{f}^T = [\,P_1, M_1, P_2, M_2\,]$ is given by the internal forces at the nodes, see (2.65). The relation $\mathsf{f} = \mathsf{K}\mathsf{u}$ between the nodal force and displacement vectors define the stiffness matrix K of a beam with distributed elastic spring support. The stiffness matrix is given in Box 2.15. The matrix is expressed in terms of non-dimensional coefficients $\psi_1, \cdots, \psi_6$, normalized in such a way that for vanishing spring stiffness $\psi_1 = \cdots = \psi_6 = 1$. The coefficients $\psi_i(\lambda\ell)$ are plotted in Fig. 2.49. It is seen that the coefficient ψ_1, representing the stiffness for a unit transverse displacement of a node without rotating it, increases approximately as $(\lambda\ell)^2$, when $\lambda\ell > 1$. This corresponds to a transverse stiffness of magnitude $12EI/\ell^3 \cdot (\lambda\ell)^2 = 6\sqrt{EI\,k}/\ell$. This form was already used in the formulas in Box 2.14 for a semi-infinite beam. It is also seen that the coefficients ψ_2, ψ_4, ψ_6, representing

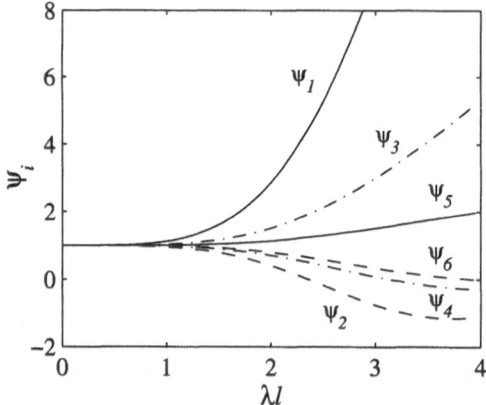

Figure 2.49: Coefficients $\psi_1, \cdots, \psi_6$ for beam on elastic springs.

the coupling between the ends of the beam do not increase in magnitude, and thus the ends effectively uncouple for $\lambda\ell > 5$ as already discussed in connection with the fluid filled tank in Example 2.25.

Extended elasticity model

The simple spring model is not well suited to represent a beam resting on an extended elastic body. The reason is, that the springs only act in response to the local displacement. This implies that according to the spring model a beam of finite length supporting a uniform load p_0 will simply obtain a uniform displacement $w = p_0/k$. As the transverse displacement is constant along the beam the internal moment and thereby also the shear force will vanish identically, and the load is transferred directly to the springs. This

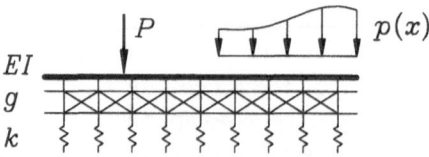

Figure 2.50: Beam on combined shear layer and spring foundation.

Figure 2.51: Spring and shear layer mechanisms.

does not correspond to either experience or theory for beams or plates resting on a soft material. The correct response should reflect that around the ends of the beam the supporting material is only loaded to one side, and therefore the settlement of the ends is less than the settlement of the central part.

A model that better represents contact with an elastic body can be obtained by introducing a layer with shear stiffness, as indicated in Fig. 2.50. If the displacements of two neighboring springs are w_i and w_{i+1}, respectively, the part of the shear layer between the springs receives the shear strain $(w_{i+1} - w_i)/\Delta x$, i.e. a shear strain proportional to the slope dw/dx. This results in a shear force of magnitude $Q_s = g\, dw/dx$. The transverse reaction acting on the beam is the derivative of this shear force, i.e. a load

$$p_s = \frac{dQ_s}{dx} = \frac{d}{dx}\left(g\frac{dw}{dx}\right) \tag{2.119}$$

The mechanisms of the springs with stiffness k and the shear layer are sketched in Fig. 2.51

The differential equation for a beam resting on the combined spring and shear layer foundation follows from substitution of the loads shown in Fig. 2.51 into the beam equation.

$$\frac{d^2}{dx^2}\left(EI\frac{d^2w}{dx^2}\right) - \frac{d}{dx}\left(g\frac{dw}{dx}\right) + k\,w(x) = p(x) \tag{2.120}$$

For constant bending, spring and shear stiffness the differential equation can be written in normalized form as

$$\frac{d^4w}{dx^4} - 4\mu^2\frac{d^2w}{dx^2} + 4\lambda^4\,w(x) = \frac{p(x)}{EI} \tag{2.121}$$

where the shear and spring stiffness are represented via the parameters μ and λ, defined by

$$\mu^2 = \frac{g}{4EI} \quad , \quad \lambda^4 = \frac{k}{4EI} \tag{2.122}$$

The parameter λ is the same as introduced above. The seemingly small change in the differential equation makes the solutions considerably more complicated.

The homogeneous equation is solved by considering partial solutions of the form $w(x) \propto \exp(\alpha x)$. Substitution of this into the normalized differential equation (2.121) leads to the characteristic equation

$$\alpha^4 - 4\mu^2\alpha^2 + 4\lambda^4 = 0 \tag{2.123}$$

For $\mu < \lambda$ the solution is complex and can be written in the convenient form

$$\alpha^2 = 2\lambda^2\left[\left(\frac{\mu}{\lambda}\right)^2 \pm i\sqrt{1 - \left(\frac{\mu}{\lambda}\right)^2}\right] \tag{2.124}$$

where i is the imaginary unit. The solution for $\mu > \lambda$ will be obtained by reformulating this form in the specific examples. α^2 is a complex number of the form

$$\alpha^2 = 2\lambda^2[\cos(2\varphi) \pm i\sin(2\varphi)] \tag{2.125}$$

where $\cos(2\varphi) = (\mu/\lambda)^2$. When written in this form the square root α can be extracted directly as

$$\alpha = \sqrt{2}\lambda[\pm\cos\varphi \pm i\sin\varphi] = \pm\sqrt{\lambda^2 + \mu^2} \pm i\sqrt{\lambda^2 - \mu^2} \tag{2.126}$$

by using standard trigonometric formulae for $\cos\varphi$ and $\sin\varphi$ in terms of $\cos(2\varphi)$. It is convenient to introduce the notation

$$\lambda_+ = \sqrt{\lambda^2 + \mu^2} \quad , \quad \lambda_- = \sqrt{\lambda^2 - \mu^2} \tag{2.127}$$

for the real and imaginary parts of α. Note, that in the absence of a shear layer $\mu = 0$, and $\lambda_+ = \lambda_- = \lambda$. Thus, the previous solutions are recovered as special cases by omitting the subscript on λ.

The complete solution to the homogeneous equation can now be written in terms of exponentials and trigonometric functions as

$$\begin{aligned} w(x) &= e^{-\lambda_+ x}\left[C_1\cos(\lambda_- x) + C_2\sin(\lambda_- x)\right] \\ &\quad + e^{\lambda_+ x}\left[C_3\cos(\lambda_- x) + C_4\sin(\lambda_- x)\right] \end{aligned} \tag{2.128}$$

This generalizes the solution (2.103) for the case of simple spring support. The solution for the case $\mu > \lambda$ is obtained from (2.128) by observing that λ_- then becomes imaginary, leading to replacement of the trigonometric functions with the corresponding hyperbolic functions, as demonstrated in the following example.

Example 2.26

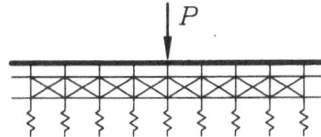

Figure 2.52: Infinite beam on combined elastic foundation.

Figure 2.52 shows a beam with bending stiffness EI supported by a generalized elastic foundation with parameters g and k. It is loaded by a concentrated force P at $x = 0$. The objective is to determine the displacement $w(x)$ and the internal moment $M(x)$. The procedure from the special case $g = 0$, treated in Example 2.23, is also followed here. Due to symmetry only the part $x > 0$ is considered explicitly.

The general solution, that remains bounded for $x \to \infty$, is of the form

$$w(x) = e^{-\lambda_+ x}\left[C_1 \cos(\lambda_- x) + C_2 \sin(\lambda_- x)\right] \;,\quad x \geq 0$$

Symmetry imposes the condition $\mathrm{d}w/\mathrm{d}x = 0$ for $x = 0$. The derivative is evaluated as

$$\frac{\mathrm{d}w}{\mathrm{d}x} = e^{-\lambda_+ x}\left[(-\lambda_+ C_1 + \lambda_- C_2)\cos(\lambda_- x) - (\lambda_+ C_2 + \lambda_- C_1)\sin(\lambda_- x)\right]$$

The condition $\mathrm{d}w/\mathrm{d}x = 0$ at $x=0$ is satisfied by selecting

$$C_1 = \frac{C}{\lambda_+} \quad,\quad C_2 = \frac{C}{\lambda_-}$$

whereby the solution takes he form

$$w(x) = C e^{-\lambda_+ x}\left[\frac{\cos(\lambda_- x)}{\lambda_+} + \frac{\sin(\lambda_- x)}{\lambda_-}\right] \;,\quad x \geq 0$$

The constant C follows by integrating the equilibrium equation from $x=0$ to infinity. The load contributes half of the force,

$$\tfrac{1}{2}P = \int_0^\infty \left(EI\frac{\mathrm{d}^4 w}{\mathrm{d}x^4} - g\frac{\mathrm{d}^2 w}{\mathrm{d}x^2} + k\,w(x)\right)\mathrm{d}x = \int_0^\infty k\,w(x)\,\mathrm{d}x$$

where the even derivatives vanish upon integration. The shear layer does not contribute to carrying the load, although it helps in distributing it. It is easily verified by differentiation that

$$\int_\infty^x w(x)\,\mathrm{d}x = -\frac{2C\,e^{-\lambda_+ x}}{\lambda_+^2 + \lambda_-^2}\left[\cos(\lambda_- x) + \frac{\lambda_+^2 - \lambda_-^2}{2\lambda_+ \lambda_-}\sin(\lambda_- x)\right]$$

This integral is substituted into the previous formula for $\frac{1}{2}P$, determining the constant

$$C = (\lambda_+^2 + \lambda_-^2)\frac{P}{4k} = \lambda^2 \frac{P}{2k}$$

The displacement then follows as

$$w(x) = \lambda^2 \frac{P}{2k} e^{-\lambda_+ x} \left[\frac{\cos(\lambda_- x)}{\lambda_+} + \frac{\sin(\lambda_- x)}{\lambda_-} \right] , \quad x \geq 0$$

The rotation is then obtained by differentiation as

$$\theta(x) = -\frac{dw}{dx} = \frac{P}{4EI} \frac{1}{\lambda_+ \lambda_-} e^{-\lambda_+ x} \sin(\lambda_- x) , \quad x \geq 0$$

and the moment by differentiation of the rotation,

$$M(x) = EI \frac{d\theta}{dx} = \frac{P}{4} e^{-\lambda_+ x} \left[\frac{\cos(\lambda_- x)}{\lambda_+} - \frac{\sin(\lambda_- x)}{\lambda_-} \right] , \quad x \geq 0$$

The solution appears as an easily recognizable generalization of the special case $g = 0$, established in Example 2.23. The effect of the shear layer can be seen in the expressions for the displacement and the moment under the force,

$$w(0) = \frac{\lambda^2}{\sqrt{\lambda^2 + \mu^2}} \frac{P}{2k} \quad , \quad M(0) = \frac{P}{4\sqrt{\lambda^2 + \mu^2}}$$

Both $w(0)$ and $M(0)$ decrease in the same way with increasing μ.

The solution in this example has been established for the case $\mu < \lambda$. However, it is a simple matter to rearrange the solution to cover the case $\mu > \lambda$. In that case the characteristic roots are

$$\alpha = \pm \lambda_+ \pm \mu_- \quad , \quad \mu_- = \sqrt{\mu^2 - \lambda^2}$$

It is noted, that the parameter λ_- only appears in the combinations $\sin(\lambda_- x)/\lambda_-$ and $\cos(\lambda_- x)$. The solution remains valid for $\mu > \lambda$, when these expressions are replaced by $\sinh(\mu_- x)/\mu_-$ and $\cosh(\mu_- x)$, respectively.

———————

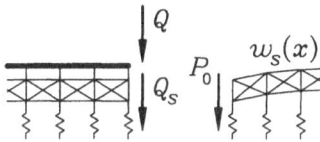

Figure 2.53: Boundary condition on combined spring and shear layer.

An essential feature of the extended elastic foundation model is the modification of the boundary conditions. In the simple spring foundation model the ends of the beam were subject to externally prescribed boundary conditions. Thus the moment and shear force would vanish, if no external loads were applied directly to the ends. This situation is changed in the extended model, because the shear layer extends continuously beyond the ends of the beam. This is illustrated in Fig. 2.53, showing the right end of a beam and the shear layer extending to infinity to the right. Introduce a coordinate x with origin at the beam end. The combined spring and shear layer outside the beam is governed by the differential equation (2.120) with vanishing bending stiffness and no surface load,

$$-\frac{\mathrm{d}}{\mathrm{d}x}\Big(g\,\frac{\mathrm{d}w_s}{\mathrm{d}x}\Big) \;+\; k\,w_s(x) \;=\; 0 \qquad , \qquad x > 0 \qquad (2.129)$$

The displacement $w_s(x)$ must vanish at infinity, and the solution then is

$$w_s(x) \;=\; w_0 \,\exp\big(-\sqrt{k/g}\,x\big) \qquad , \qquad x > 0 \qquad (2.130)$$

The displacement of the combined spring shear layer for $x > 0$ requires a concentrated downward force P_0 at $x=0$. This force is determined from the shear stiffness of the layer as

$$P_0 \;=\; -\,g\,\frac{\mathrm{d}w_s}{\mathrm{d}x}\Big|_{x=0} \;=\; \sqrt{k\,g}\,w_0 \qquad (2.131)$$

Note that P_0 vanishes for a simple spring layer.

While the shear layer outside the beam is loaded by the concentrated force P_0, the end of the beam has the shear force Q and the shear layer under the beam has the shear force Q_s. Transverse equilibrium at the end of the beam requires the sum of these forces to vanish, whereby the shear force condition on the right end of the beam takes the form

$$Q \;+\; g\,\frac{\mathrm{d}w}{\mathrm{d}x} \;+\; \sqrt{k\,g}\,w_0 \;=\; 0 \qquad (2.132)$$

When the shear force in the beam is expressed in terms of the third displacement derivative, the boundary condition becomes

$$EI\frac{d^3w}{dx^3} - g\frac{dw}{dx} - \sqrt{k\,g}\,w = 0 \qquad (2.133)$$

This boundary can be expressed in terms of the non-dimensional parameters μ and λ as

$$\frac{d^3w}{dx^3} - 4\mu^2\frac{dw}{dx} - 4\mu\,\lambda^2\,w = 0 \qquad (2.134)$$

This is the boundary condition to be used for no transverse load at the right end of the beam. At the left end of the beam the sign of the last term should be changed. The zero moment condition is the same as previously, i.e. $d^2w/dx^2 = 0$.

The boundary condition (2.134) is an essential part of the combined spring shear layer model, because it introduces a coupling to the part of the layer outside the beam. However, the rather complicated form of this boundary condition in combination with the non-trivial form of the general solution (2.128) makes it difficult to derive manageable closed form solutions. The following example shows, how the problem of a uniformly loaded beam of finite length can be solved by careful selection of the form of the solution.

Example 2.27

Consider the uniformly loaded beam of length ℓ in a combined spring shear layer foundation, shown in Fig. 2.54. A coordinate x is introduced with origin at the left end of the beam, and in accordance with the previous discussion of spring supported beams of finite length a complementary coordinate $x' = \ell - x$ is introduced.

The problem has a symmetric solution and therefore includes two arbitrary constants to match a zero moment condition at the ends and the

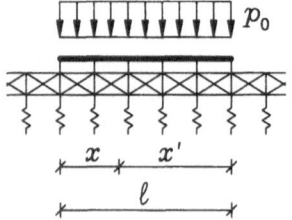

Figure 2.54: Uniformly loaded beam on combined foundation.

shear condition (2.134). If the solution can be formulated initially in such a way that it satisfies the zero moment condition, it will contain only a single arbitrary constant to be determined by substitution into (2.134). This can be obtained by observing that, if the original beam equation (2.120) is differentiated twice, it is seen that the second derivative d^2w/dx^2 satisfies the original homogeneous equation. A symmetric solution vanishing at both ends will therefore be a suitable representation of d^2w/dx^2, and thereby of the internal moment distribution $M(x)$. This representation of $M(x)$ is obtained by combining a hyperbolic function of λ_+x' with a trigonometric function of λ_-x. The corresponding symmetric function is

$$\frac{d^2w}{dx^2} = C\left[\sinh(\lambda_+x')\sin(\lambda_-x) + \sinh(\lambda_+x)\sin(\lambda_-x')\right]$$

The arbitrary constant C must be determined to satisfy the boundary condition (2.134). This boundary condition involves the displacement distribution $w(x)$ as well as its first and third derivatives.

The third derivative follows directly by differentiation.

$$\frac{d^3w}{dx^3} = C\left[-\lambda_+\Big(\cosh(\lambda_+x')\sin(\lambda_-x) - \cosh(\lambda_+x)\sin(\lambda_-x')\Big)\right.$$
$$\left.+ \lambda_-\Big(\sinh(\lambda_+x')\cos(\lambda_-x) - \sinh(\lambda_+x)\cos(\lambda_-x')\Big)\right]$$

The fourth derivative is obtained by further differentiation.

$$\frac{d^4w}{dx^4} = 2C\left[\mu^2\Big(\sinh(\lambda_+x')\sin(\lambda_-x) + \sinh(\lambda_+x)\sin(\lambda_-x')\Big)\right.$$
$$\left.- \lambda_+\lambda_-\Big(\cosh(\lambda_+x')\cos(\lambda_-x) + \cosh(\lambda_+x)\cos(\lambda_-x')\Big)\right]$$

The displacement distribution $w(x)$ can now be determined from the beam differential equation (2.121), including the load $p(x) = p_0$.

$$w = \frac{p_0}{k} + \frac{\mu^2}{\lambda^4}\frac{d^2w}{dx^2} - \frac{1}{4\lambda^4}\frac{d^4w}{dx^4}$$
$$= \frac{p_0}{k} + \frac{C}{2\lambda^4}\left[\mu^2\Big(\sinh(\lambda_+x')\sin(\lambda_-x) + \sinh(\lambda_+x)\sin(\lambda_-x')\Big)\right.$$
$$\left.+ \lambda_+\lambda_-\Big(\cosh(\lambda_+x')\cos(\lambda_-x) + \cosh(\lambda_+x)\cos(\lambda_-x')\Big)\right]$$

This expression is then differentiated to obtain the first derivative. In the present case the load is constant and therefore does not contribute to the derivative.

$$\frac{dw}{dx} = -\frac{C}{2\lambda^2}\Big[\lambda_+\big(\cosh(\lambda_+x')\sin(\lambda_-x) - \cosh(\lambda_+x)\sin(\lambda_-x')\big)$$
$$+ \lambda_-\big(\sinh(\lambda_+x')\cos(\lambda_-x) - \sinh(\lambda_+x)\cos(\lambda_-x')\big)\Big]$$

The displacement distribution $w(x)$ and its first four derivatives have now been determined in terms of the arbitrary constant C.

The constant C is determined by substitution into the shear boundary condition at $x = \ell$. Here $x' = 0$, and the boundary condition determines the constant as

$$C = -2\mu\lambda^2\frac{p_0}{k}\Big[\lambda_-\Big(1 + 2\frac{\mu^2}{\lambda^2}\Big)\sinh(\lambda_+\ell) + \lambda_+\Big(1 - 2\frac{\mu^2}{\lambda^2}\Big)\sin(\lambda_-\ell)$$
$$+ 2\mu\frac{\lambda_+\lambda_-}{\lambda^2}\big(\cosh(\lambda_+\ell) + \cos(\lambda_-\ell)\big)\Big]^{-1}$$

The full solution has now been determined. It is seen that the expression for the constant C contains the factor μ, and thus vanishes for a simple spring foundation, leaving the special solution $w(x) = p_0/k$.

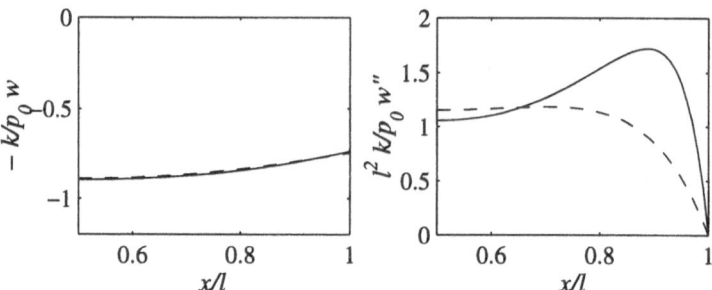

Figure 2.55: Displacement and curvature, $E/E_s = 200$ (−), 2000 (- -).

Figure 2.55 shows the displacement $w(x)$ and the curvature, representing the moment $M(x)$, for the right half of a beam on a composite foundation. The beam is of length $\ell = 4\,\text{m}$, with cross-section height $h = 0.1\,\text{m}$ and width $b = 0.1\,\text{m}$. The composite elastic foundation is modeled as an elastic layer with depth $H = \ell$, Poisson ratio $\nu_s = 0.25$ and elastic modulus E_s. By the theory presented below the composite layer parameters are determined from (2.138) as $k = 1.20E_sb/H$ and $g = 0.12E_sbH$. Two cases are considered: a flexible beam with $E/E_s = 200$, and a stiff beam with $E/E_s = 2000$. The displacement is given in the normalized form

$w(x)k/p_0$. If there had not been any shear layer in the foundation, this normalized displacement would be identically equal to one. However, the composite foundation leads to a concentrated force P_0 at each end of the beam, equal to the part of the load carried by the part of the foundation outside the beam as shown in Fig. 2.53. This reduces the displacement, as seen in Fig. 2.55a. The end forces also lead to bending in the beam. The distribution of the bending moment is illustrated via the normalized curvature $(d^2w/dx^2)\ell^2 k/p_0$. It is seen, that the flexible beam has local curvature maxima close to the ends of the beam, while the curvature of the stiff beam increases monotonically towards the center of the beam.

The solution method presented in this example closely resembles the technique of introducing the functions $A_*(\lambda x), B_*(\lambda x), \cdots$ defined by (2.113) for finite beams on a simple spring foundation. Also in that case the key feature was the homogeneous boundary conditions of the function $B_*(\lambda x)$. The differentiation rules of these functions can be used to check the formulae of this example for $\mu = 0$.

In order to obtain realistic solutions to problems involving beams on an elastic medium it is necessary to obtain a representation of the equivalent layer parameters k and g in terms of the properties of the elastic medium with modulus of elasticity E_s and Poisson ratio ν_s. This problem was originally treated by Vlasov, and closed form results later obtained by Vallabhan & Das (1991). It turns out that the optimal values of the parameters k and g depend not only on the elastic parameters E_s, ν_s, but also on the depth H of the elastic medium as well as the beam displacement distribution, see Fig. 2.56. The optimal parameters are determined by assuming that the displacement field (u, w) in the elastic medium can be approximately represented by a vertical motion $w(x, z) = w_s(x)\varphi(z)$ in product format. The procedure involves an argument about the elastic energy of the medium outside the scope of this book, but clearly presented by Vallabhan & Das (1991).

The result of this analysis is that the depth variation $\varphi(z)$ is governed by a parameter γ, determined by

$$\left(\frac{\gamma}{H}\right)^2 = \frac{1 - 2\nu_s}{2(1 - \nu_s)} \frac{\int_{-\infty}^{\infty}\left(\frac{dw_s}{dx}\right)^2 dx}{\int_{-\infty}^{\infty} w_s^2 \, dx} \tag{2.135}$$

Outside the beam the surface displacement w_s is described by the exponential

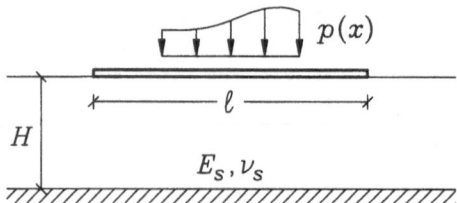

Figure 2.56: Beam on layer with elastic parameters E_s and ν_s.

variation (2.130). Thus the formula (2.135) can be written as

$$\left(\frac{\gamma}{H}\right)^2 = \frac{1 - 2\nu_s}{2(1 - \nu_s)} \frac{\int_0^\ell \left(\frac{\mathrm{d}w}{\mathrm{d}x}\right)^2 \mathrm{d}x + \frac{1}{2}\sqrt{k/g}\left[w_0^2 + w_\ell^2\right]}{\int_0^\ell w^2\,\mathrm{d}x + \frac{1}{2}\sqrt{g/k}\left[w_0^2 + w_\ell^2\right]} \tag{2.136}$$

The general expression depends on the specific deformation of the beam, but for relatively stiff beams the variation of the displacement is nearly constant under the beam, leading to the following approximation for symmetrically loaded stiff beams

$$\left(\frac{\gamma}{H}\right)^2 \simeq \frac{1 - 2\nu_s}{2(1 - \nu_s)} \frac{\sqrt{k/g}}{\ell + \sqrt{g/k}} \tag{2.137}$$

For a given value of the parameter γ the spring and shear layer parameters are determined by

$$k = \frac{(1 - \nu_s)\,E_s}{(1 + \nu_s)(1 - 2\nu_s)} \frac{b}{H} \psi_k(\gamma) \quad , \quad g = \frac{E_s\,b\,H}{2(1 + \nu_s)} \psi_g(\gamma) \tag{2.138}$$

where b is the width of the beam, and the functions $\psi_k(\gamma)$ and $\psi_g(\gamma)$ are

$$\psi_k(\gamma) = \gamma\frac{\sinh\gamma\cosh\gamma + \gamma}{2\sinh^2\gamma} \quad , \quad \psi_g(\gamma) = \frac{1}{\gamma}\frac{\sinh\gamma\cosh\gamma - \gamma}{2\sinh^2\gamma} \tag{2.139}$$

These functions are shown in Fig. 2.57 for the relevant range of γ.

If the formulae (2.138) and (2.139) are used to express the ratio k/g, the foundation depth to beam length ratio H/ℓ can be expressed as a function of the parameter γ by use of (2.137). After some reduction the following result is obtained,

$$\frac{H}{\ell} = \sqrt{\frac{2(1 - \nu_s)}{1 - 2\nu_s}} \frac{1}{4}\sqrt{\sinh^2(2\gamma) - (2\gamma)^2} \tag{2.140}$$

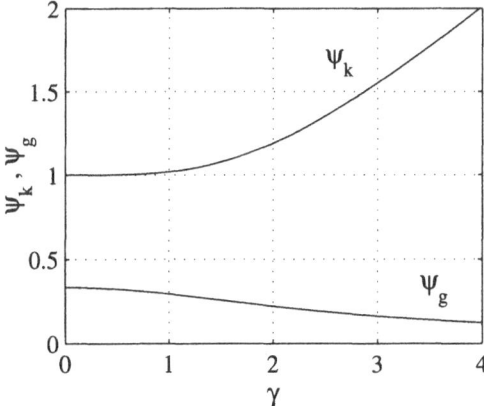

Figure 2.57: Functions $\psi_f(\gamma)$ and $\psi_g(\gamma)$ for the parameters k and g.

Figure 2.58 shows the parameter γ as function of H/ℓ for a stiff beam resting on an elastic layer with $\nu_s = 0.25$. The parameter γ increases with foundation depth, but the increase is moderate, and typical values of γ are in the range 0.5–3. This in turn implies that the coefficint ψ_k is of the order 1-1.5, and ψ_g around 0.2–0.3. In view of the uncertainties of foundation properties these approximate values can be used directly in (2.138). The parameters used in Example 2.27 correspond to $H/\ell \simeq 1$ and $\gamma \simeq 1$.

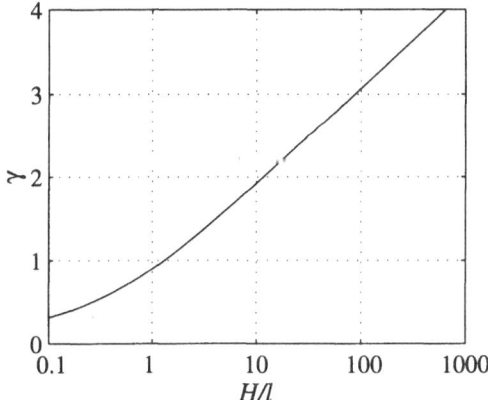

Figure 2.58: Parameter γ as function of H/ℓ for stiff beam and $\nu_s = 0.25$.

2.9   Inertial forces and beam vibrations

According to Newton's second law acceleration of a body requires a force that is proportional to the mass of the body. In a deformable body it is necessary to account for the acceleration of each part of the body. The equations of motion for a beam are obtained by including the acceleration of each part of the beam in the force and moment balance equations. Figure 2.59 shows the forces and moments acting on a thin slice of the beam of thickness dx. The same figure was used in Section 2.2 to derive the equilibrium equations of a beam. The equations of motion are obtained by considering the forces necessary to produce the transverse acceleration $\partial^2 w/\partial t^2$ of the beam section with area A and mass density ρ. It is customary to neglect the inertial effects associated with rotation of the beam section, and it can be shown that this is a higher order correction for slender beams, see e.g. Inman (1996).

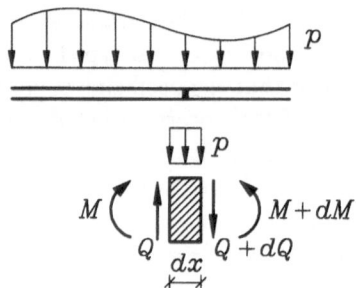

Figure 2.59: Equations of motion of thin beam section.

The total transverse force acting on the beam section of thickness dx is $dQ + p\,dx$. This force gives the mass $\rho A\,dx$ the acceleration $\partial^2 w/\partial t^2$. When this statement is normalized by the section thickness dx, the transverse equation of motion is obtained in the form

$$\frac{\partial Q}{\partial x} + p = \rho A \frac{\partial^2 w}{\partial t^2} \qquad (2.141)$$

If the inertial effects of rotating the section are neglected, the moment balance equation is

$$\frac{\partial M}{\partial x} = Q \qquad (2.142)$$

Elimination of the shear force Q between these two equations leads to the equation of motion in the form

$$\frac{\partial^2 M}{\partial x^2} - \rho A \frac{\partial^2 w}{\partial t^2} + p = 0 \tag{2.143}$$

In these equations the symbol ∂ has been introduced to denote partial differentiation, because the displacement and internal forces are now functions of the two independent variables x and t.

The equation of motion (2.143) generalizes the static equilibrium equation (2.15). The dynamic effect of accelerating the mass may be looked at as an additional *inertial force* $-\rho A \partial^2 w/\partial t^2$ to be added to the external force $p(x,t)$. This approach, in which inertial forces are included among the static forces in the equilibrium equations, is called d'Alembert's principle. It enables direct formulation of dynamic equations of motion from the static equilibrium equations for many types of structures like beams, cables, plates and shells.

The equation of motion (2.143) contains derivatives of the moment M and the displacement w. In order to solve a dynamic beam problem this equation must be supplemented with a constitutive equation that relates the moment M to the deformation of the beam. This relation was derived for linearized beam bending theory in Sections 2.1 and 2.2 in the form

$$M = -EI \frac{\partial^2 w}{\partial x^2} \tag{2.144}$$

When the moment is eliminated between the equation of motion (2.143) and the constitutive equation (2.144), the resulting dynamic beam equation is

$$\frac{\partial^2}{\partial x^2}\left(EI\frac{\partial^2 w}{\partial x^2}\right) + \rho A \frac{\partial^2 w}{\partial t^2} = p \tag{2.145}$$

This equation can model two types of seemingly different dynamic phenomena in beams, wave propagation and vibrations. While beam vibrations is the main theme of this section, a brief discussion of wave propagation in beams is given in the following example.

Example 2.28

Consider the idealized problem of an infinitely long homogeneous beam without external load. It is of theoretical as well as practical interest to investigate the possible existence of wave propagation along the beam. The motion of the beam is governed by the partial differential equation

$$\frac{\partial^4 w}{\partial x^4} + \frac{\rho A}{EI}\frac{\partial^2 w}{\partial t^2} = 0$$

It turns out that this equation permits solutions of the form

$$w(x,t) = w_0 \sin(\omega t - \beta x)$$

for suitable choice of the parameters k and ω. In this solution the independent variables x and t only appear in the combination

$$\varphi = \omega t - \beta x$$

called the *phase angle*.

The parameters ω and β can be given a physical interpretation in the following way. Consider a fixed point $x = x_0$. At that particular point the solution has the form

$$w(x_0, t) = w_0 \sin(\omega t - \varphi_0) \quad , \quad \varphi_0 = \beta x_0$$

This is a harmonic vibration with *angular frequency* ω. The motion is periodic with period T. The period corresponds to an increase of the phase angle by 2π, and thus the period and the angular frequency are related by

$$\omega = \frac{2\pi}{T}$$

Similarly the solution can be considered at a particular time, e.g. $t = 0$. At this time the solution appears as a sine-wave. The parameter β is called the wave number, and the wave length λ corresponds to an increase of the phase angle of 2π. Thus the wave number and the wave length are related by

$$\beta = \frac{2\pi}{\lambda}$$

The solution is shown for $t_0 = 0$ and $t_1 > t_0$ in Fig. 2.60.

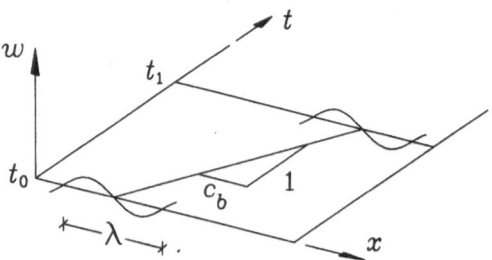

Figure 2.60: Propagation of harmonic wave in an infinite beam.

It is easily seen that points (x_0, t_0) and $(x_0 + \Delta x, t_0 + \Delta t)$ have the same phase angle, if the space and time increments satisfy the relation

$$c_b = \frac{\Delta x}{\Delta t} = \frac{\omega}{\beta} = \frac{\lambda}{T}$$

Thus, any particular value of the phase angle will appear to be traveling towards increasing values of x with the *wave speed* $c_b = \lambda/T$, where the subscript b indicates bending waves. The wave speed c_b is found by substituting the harmonic wave solution into the homogeneous equation of motion of the beam. This gives

$$\frac{\omega^2}{\beta^4} = \frac{EI}{\rho A}$$

This relation determines the speed of bending waves as

$$c_b = \frac{\omega}{\beta} = 2\pi \sqrt{\frac{I}{A\lambda^2}} \sqrt{\frac{E}{\rho}}$$

In this formula $c_l = \sqrt{E/\rho}$ is the speed of longitudinal waves in the beam, and the bending moment of inertia of a rectangular beam was given in Example 2.1 as $I = \frac{1}{12} Ah^2$. The speed of bending waves in a beam with rectangular cross-section can then be expressed in the form

$$c_b = \frac{2\pi}{\sqrt{12}} \frac{h}{\lambda} \sqrt{\frac{E}{\rho}} \simeq 1.8 \frac{h}{\lambda} c_l$$

It is seen that the speed of bending waves decreases with increasing wave length. This implies that the individual harmonic components of any particular non-harmonic wave pattern will propagate with different speed, and thus the wave form will change with time. This type of wave behaviour is called dispersive, because a local wave phenomenon will disperse along the x-axis. The simple beam bending theory presented here is based on the assumption that $\lambda \gg h$, and thus bending waves have lower wave speed than longitudinal waves.

Most dynamic beam problems involve beams of finite length, and thus a proper formulation of the problem requires definition of appropriate *boundary conditions* at the ends of the beam, and *initial conditions* defining the state of each point of the beam at some initial time t_0, often taken as $t_0 = 0$. A

Box 2.16: Differential equation of beam vibrations

The dynamics of Bernoulli beams is described by the differential equation of motion

$$\frac{\partial^2}{\partial x^2}\left(EI\frac{\partial^2 w}{\partial x^2}\right) + \rho A \frac{\partial^2 w}{\partial t^2} = p(x,t)$$

supplemented with boundary conditions, formulated in terms of

$$w \quad , \quad \frac{\partial w}{\partial x} \quad , \quad M = -EI\frac{\partial^2 w}{\partial x^2} \quad , \quad Q = -\frac{\partial}{\partial x}\left(EI\frac{\partial^2 w}{\partial x^2}\right)$$

and initial conditions, prescribing displacement and velocity at the initial time t_0,

$$w(x,t_0) = w_0(x) \quad , \quad \frac{\partial w}{\partial t}(x,t_0) = \dot{w}_0(x)$$

Solutions to beam vibration problems are typically obtained by the method of separation of variables.

---

typical problem is illustrated in Fig. 2.61, showing a beam of length ℓ with a fixed end at $x = 0$ and a free end at $x = \ell$. The behaviour at the ends of the beam must be specified in terms of displacement, rotation, moment or shear force. Initial conditions define the position and the velocity for all cross-sections of the beam as indicated in Box 2.16. In the following it will be

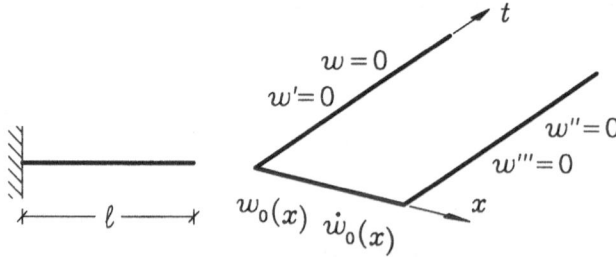

Figure 2.61: Boundary and initial conditions of beam.

convenient to denote spatial partial derivatives as $\partial w/\partial x = x'$ and temporal partial derivatives as $\partial w/\partial t = \dot{x}$. This notation is used in Fig. 2.61.

A free harmonic vibration with angular frequency ω can be represented in the form

$$w(x,t) \;=\; \tilde{w}(x)\cos\left(\omega t - \varphi_0\right) \tag{2.146}$$

Substitution of this into the equation of motion (2.145) with $p(x,t) = 0$ gives the ordinary differential equation

$$\frac{\mathrm{d}^2}{\mathrm{d}x^2}\left(EI\frac{\mathrm{d}^2\tilde{w}}{\mathrm{d}x^2}\right) \;-\; \omega^2\rho A\,\tilde{w} \;=\; 0 \tag{2.147}$$

For a homogeneous beam this equation can be written as

$$\frac{\mathrm{d}^4\tilde{w}}{\mathrm{d}x^4} \;-\; \beta^4\,\tilde{w} \;=\; 0 \tag{2.148}$$

with the parameter β given as

$$\beta^4 \;=\; \omega^2\,\frac{\rho A}{EI} \tag{2.149}$$

It is interesting to observe, that this equation is similar to (2.98) for a beam on a simple spring support, but with the notable difference that in the present case the spring constant is negative, $k = -\omega^2\rho A$. This change of sign has a profound influence on the solution.

The general solution to (2.148) can be expressed in terms of trigonometric and hyperbolic functions as

$$\tilde{w}(x) \;=\; C_1\sin(\beta x) + C_2\cos(\beta x) + C_3\sinh(\beta x) + C_4\cosh(\beta x) \tag{2.150}$$

In the case of free vibrations the boundary conditions are homogeneous. A non-trivial solution can therefore only be obtained for particular values of the parameter β. The solution procedure therefore consists in the determination of these values $\beta_1, \beta_2, \cdots$, and the corresponding values of the arbitrary constants $C_1, \cdots, C_4$. These solutions represent the mode-shapes of free vibrations with frequencies $\omega_1, \omega_2, \cdots$ determined from (2.149). The procedure is illustrated in the following simple example.

Example 2.29

Consider free vibrations of a simply supported beam of length ℓ. The boundary conditions are $w = 0$, $w'' = 0$ at $x = 0$ and $x = \ell$. From (2.150) the spatial solutions are found to be

$$\tilde{w}_n(x) = C_n \sin(\beta_n x) = C_n \sin\left(\frac{n\pi}{\ell}x\right) \quad , \quad n = 1, 2, \cdots$$

with the parameters $\beta_n = n\pi/\ell$. The angular frequency for each of these vibration modes are found from (2.149)

$$\omega_n = \beta_n^2 \sqrt{\frac{EI}{\rho A}} = \left(\frac{n\pi}{\ell}\right)^2 \sqrt{\frac{EI}{\rho A}} \quad , \quad n = 1, 2, \cdots$$

Note, that while the wave number β_n is proportional to n, the angular frequency ω_n is proportional to n^2.

The general free vibration solution of the simply supported beam is represented as a sum of the individual vibration modes in the form

$$w(x, t) = \sum_{n=1}^{\infty} \left[A_n \cos(\omega_n t) + B_n \sin(\omega_n t) \right] \sin(\beta_n x)$$

The arbitrary constants A_n and B_n are to be determined from the initial conditions,

$$w(x, 0) = \sum_n A_n \sin(\beta_n) = w_0(x)$$
$$\dot{w}(x, 0) = \sum_n \omega_n B_n \sin(\beta_n) = \dot{w}_0(x)$$

The arbitrary constants A_n and B_n are seen to appear as coefficients in a Fourier sine series expansion of $w_0(x)$ and $\dot{w}_0(x)$, respectively. These coefficients are determined by use of the orthogonality relation

$$\frac{2}{\ell} \int_0^\ell \sin\left(\frac{n\pi}{\ell}x\right) \sin\left(\frac{m\pi}{\ell}x\right) dx = \begin{cases} 1 & \text{for} \quad m = n \\ 0 & \text{for} \quad m \neq n \end{cases}$$

Multiplication with $\sin(\beta_m x)$, followed by integration, then gives the arbitrary constants

$$A_n = \frac{2}{\ell} \int_0^\ell \sin(\beta_n x)\, w_0(x)\, dx \quad , \quad \omega_n B_n = \frac{2}{\ell} \int_0^\ell \sin(\beta_n x)\, \dot{w}_0(x)\, dx$$

This determines the full free vibration solution in terms of the frequencies ω_n, the corresponding mode shapes $\sin(\beta_n x)$, and the initial displacement $w_0(x)$ and velocity $\dot{w}_0(x)$ distributions.

The free vibration solution in terms of mode shapes appears quite different from the wave solution considered in Example 2.28. However, this is mainly a question of grouping of the terms. The products of functions of time and space can be reformulated by use of the trigonometric product formulae

$$2\cos(\omega_n t)\sin(\beta_n x) = -\sin(\omega_n t - \beta_n x) + \sin(\omega_n t + \beta_n x)$$

$$2\sin(\omega_n t)\sin(\beta_n x) = \cos(\omega_n t - \beta_n x) - \cos(\omega_n t + \beta_n x)$$

When these results are introduced into the solution it takes the following form

$$w(x,t) = \sum_n \tfrac{1}{2}A_n\left[-\sin(\omega_n t - \beta_n x)) + \sin(\omega_n t + \beta_n x))\right]$$
$$+ \sum_n \tfrac{1}{2}B_n\left[\cos(\omega_n t - \beta_n x)) - \cos(\omega_n t + \beta_n x))\right]$$

This is a representation of the free vibration solution in terms of traveling waves, like those analyzed in Example 2.28. Each square bracket contains two terms: the first representing a wave traveling to the right and the second representing a wave traveling to the left. The reason for using vibration modes - or standing waves - is that they are more easily related to general boundary conditions.

In the analysis of beams with distributed elastic support it was found to be advantageous to represent the solution in terms of special combinations of trigonometric and hyperbolic functions, see e.g. (2.113). In a similar way it often leads to simpler expressions if the solution to harmonic beam vibration problems are expressed in terms of the functions

$$S_+(\beta x) = \sinh(\beta x) + \sin(\beta x) \quad, \quad C_+(\beta x) = \cosh(\beta x) + \cos(\beta x)$$

$$S_-(\beta x) = \sinh(\beta x) - \sin(\beta x) \quad, \quad C_-(\beta x) = \cosh(\beta x) - \cos(\beta x)$$
$$\tag{2.151}$$

These functions satisfy the circular differentiation rules

$$\frac{dS_+(\beta x)}{dx} = \beta\, C_+(\beta x) \quad, \quad \frac{dC_+(\beta x)}{dx} = \beta\, S_-(\beta x)$$
$$\frac{dS_-(\beta x)}{dx} = \beta\, C_-(\beta x) \quad, \quad \frac{dC_-(\beta x)}{dx} = \beta\, S_+(\beta x)$$
$$\tag{2.152}$$

It also observed, that only the function $C_+(\beta x)$ does not vanish at $x = 0$.

The general solution to the homogeneous equation (2.147) for harmonic vibrations of a beam can now be expressed in terms of these functions as

$$\tilde{w}(x) \;=\; C_1\,S_+(\beta x) + C_2\,C_+(\beta x) + C_3\,S_-(\beta x) + C_4\,C_-(\beta x) \qquad (2.153)$$

instead of the previous form (2.150). The differentiation rules (2.152) and the simple values at $x = 0$ makes this representation particularly convenient as illustrated in the following example.

Example 2.30

In this example the frequencies and mode shapes for the fixed–free beam of Fig. 2.61 are obtained by use of the representation (2.153). The boundary conditions at the fixed end, $x = 0$, are

$$\tilde{w}(0) \;=\; 0 \qquad , \qquad \tilde{w}'(0) \;=\; 0$$

From these conditions it follows immediately that $C_2 = 0$ and $C_1 = 0$. Thus the problem to be solved consists of the function

$$\tilde{w}(x) \;=\; C_3\,S_-(\beta x) + C_4\,C_-(\beta x)$$

with the free-end boundary conditions

$$\tilde{w}''(\ell) \;=\; 0 \qquad , \qquad \tilde{w}'''(\ell) \;=\; 0$$

Substitution of the representation of the function $\tilde{w}(x)$ into these conditions leads to the equation system

$$\begin{bmatrix} S_+(\beta\ell) & C_+(\beta\ell) \\ C_+(\beta\ell) & S_-(\beta\ell) \end{bmatrix} \begin{bmatrix} C_3 \\ C_4 \end{bmatrix} = \begin{bmatrix} 0 \\ 0 \end{bmatrix}$$

This equation system will only permit a non-trivial solution, if the determinant vanishes. Thus, the parameter β must satisfy the equation

$$S_+(\beta\ell)S_-(\beta\ell) - C_+^2(\beta\ell) \;=\; 0$$

Substitution of the hyperbolic and trigonometric functions from (2.151) gives

$$\left[\sinh^2(\beta\ell) - \sin^2(\beta\ell)\right] - \left[\cosh^2(\beta\ell) + \cos^2(\beta\ell) + 2\cosh(\beta\ell)\cos(\beta\ell)\right] = 0$$

After use of standard identities the following transcendental equation for $\beta\ell$ is obtained.

$$\cos(\beta\ell)\cosh(\beta\ell) + 1 \;=\; 0$$

It is convenient to write the equation in the form

$$\cos \beta \ell = \frac{-1}{\cosh(\beta \ell)}$$

The solution is the intersection of the graphs of the functions defined by the left and right hand sides of the equation, as illustrated in Fig. 2.62.

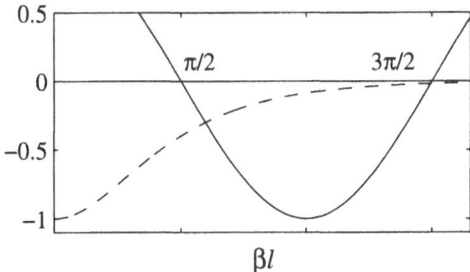

Figure 2.62: Frequency equation: $\cos(\beta \ell)$ (–) and $-1/\cosh(\beta \ell)$ (- -).

The roots of the frequency equation are given by the abscissae of the points of intersection, $\beta_1 \ell, \beta_2 \ell, \cdots$. These abscissae can be found by iteration, starting from

$$\beta_n^{(0)} \ell = (2n-1)\frac{\pi}{2} \quad , \quad n = 1, 2, \cdots$$

The iteration procedure can be formulated as

$$\beta_n^{(i+1)} \ell = (2n-1)\frac{\pi}{2} - (-1)^n \sin^{-1} \left(\frac{1}{\cosh(\beta_n^{(i)} \ell)} \right)$$

The first correction gives the approximate solution

$$\beta_n^{(1)} \ell = (2n-1)\frac{\pi}{2} - \frac{(-1)^n}{\cosh\left((n - \frac{1}{2})\pi\right)}$$

The two first steps and the final value of the wave numbers $\beta_n \ell$ are given below, together with the angular frequency, determined from (2.149).

n	1	2	3	4
$\beta_n^{(0)} \ell$	1.5708	4.7124	7.8540	10.9956
$\beta_n^{(1)} \ell$	1.9693	4.6944	7.8548	10.9955
$\beta_n \ell$	1.8751	4.6941	7.8548	10.9955
$\omega_n / (EI/\rho A \ell^4)^{1/2}$	3.5160	22.034	61.698	120.901

The table illustrates that, apart from the first two to three values of the wave numbers, the remaining wave numbers are regularly spaced and can be determined with great accuracy without iteration.

Stiffness matrix for harmonic motion

In the analysis of harmonic vibrations of beam structures it is of interest to know the relation between the nodal force vector f and the nodal displacement vector u for a homogeneous beam.

$$f^T = [P_1, M_1, P_2, M_2] \quad , \quad u^T = [w_1, \theta_1, w_2, \theta_2] \qquad (2.154)$$

The nodal force and displacement components exhibit harmonic variation in time with angular frequency ω, i.e. $P_1 = \tilde{P}_1 \cos(\omega t)$ etc. . The following equations relate nodal forces and displacements at one particular time.

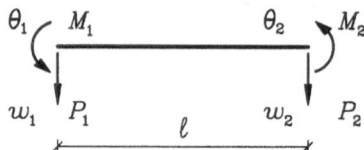

Figure 2.63: Homogeneous beam with end loads (P_1, M_1, P_2, M_2).

The nodal force and displacement components are shown in Fig. 2.63 for a particular time. They are related by

$$f = Ku \qquad (2.155)$$

where the stiffness matrix K now depends on the angular frequency ω. The static case, treated in Section 2.6, corresponds to $\omega = 0$.

The columns of the stiffness matrix consists of the nodal force vector corresponding to each of the special solutions illustrated in Fig. 2.64. The stiffness matrix is obtained by determining the shape functions $N_1(x), \cdots, N_4(x)$, illustrated in Fig. 2.64, and finding the corresponding end forces and moments. The procedure follows that of Section 2.6 with the important difference that the shape functions now satisfy the equation of motion (2.147) instead of the equilibrium equation (2.56).

Figure 2.64: Shape functions for imposed unit end motions $w_1, \theta_1, w_2, \theta_2$.

The shape functions $N_1(x), \cdots, N_4(x)$ satisfy the homogeneous equation for harmonic beam vibrations,

$$\frac{\mathrm{d}^4 N_j}{\mathrm{d}x^4} - \beta^4 N_j = 0 \quad , \quad j = 1, 2, 3, 4 \tag{2.156}$$

where the wave number β is given in terms of the angular frequency by (2.149). It is convenient to represent the shape functions in terms of the special function combinations introduced in (2.151). The shape functions $N_3(x)$ and $N_4(x)$ satisfy the homogeneous boundary conditions $N_j(0) = 0$ and $N_j'(0) = 0$, and they can therefore be expressed in terms of the functions $S_-(\beta x)$ and $C_-(\beta x)$ as

$$N_j(x) = A_j \, S_-(\beta x) + B_j \, C_-(\beta x) \quad , \quad j = 3, 4 \tag{2.157}$$

The shape functions $N_1(x)$ and $N_2(x)$ satisfy boundary conditions similar to those of $N_3(x)$ and $N_4(x)$, when the roles of the two end points are interchanged. In the following the shape functions $N_3(x)$ and $N_4(x)$ are derived in detail. The remaining functions $N_1(x)$ and $N_2(x)$ then follow by symmetry considerations.

The shape function $N_3(x)$ is represented as

$$N_3(x) = A_3 \, S_-(\beta x) + B_3 \, C_-(\beta x) \tag{2.158}$$

where the arbitrary constants A_3 and B_3 are determined from the boundary conditions at $x = \ell$,

$$N_3(\ell) = 1 \quad , \quad N_3'(\ell) = 0 \tag{2.159}$$

These boundary conditions give the equations

$$\begin{bmatrix} S_-(\beta\ell) & C_-(\beta\ell) \\ C_-(\beta\ell) & S_+(\beta\ell) \end{bmatrix} \begin{bmatrix} A_3 \\ B_3 \end{bmatrix} = \begin{bmatrix} 1 \\ 0 \end{bmatrix} \tag{2.160}$$

where the differentiation rules (2.152) have been used. The determinant of
these equations is

$$D(\beta\ell) = 2\left[\cosh(\beta\ell)\cos(\beta\ell) - 1\right] \tag{2.161}$$

and the solution then follows as

$$A_3 = \frac{S_+(\beta\ell)}{D(\beta\ell)} \quad , \quad B_3 = -\frac{C_-(\beta\ell)}{D(\beta\ell)} \tag{2.162}$$

This determines the shape function $N_3(x)$ as

$$N_3(x) = \frac{1}{D(\beta\ell)}\left[S_+(\beta\ell)\,S_-(\beta x) - C_-(\beta\ell)\,C_-(\beta x)\right] \tag{2.163}$$

This form of the solution illustrates the convenience of using the special
function combinations defined in (2.151).

The shape function $N_4(x)$ is found in a completely similar way from the
representation

$$N_4(x) = A_4\,S_-(\beta x) + B_4\,C_-(\beta x) \tag{2.164}$$

The arbitrary constants A_4 and B_4 are determined from the boundary conditions

$$N_4(\ell) = 0 \quad , \quad N_4'(\ell) = -1 \tag{2.165}$$

where the latter corresponds to the unit rotation $\theta_2 = 1$. The corresponding
equation system is

$$\begin{bmatrix} S_-(\beta\ell) & C_-(\beta\ell) \\ C_-(\beta\ell) & S_+(\beta\ell) \end{bmatrix}\begin{bmatrix} A_4 \\ B_4 \end{bmatrix} = \begin{bmatrix} 0 \\ -1/\beta \end{bmatrix} \tag{2.166}$$

where only the right side has changed. Thus the determinant is $D(\beta\ell)$, and
the arbitrary constants are

$$A_4 = \frac{1}{\beta}\frac{C_-(\beta\ell)}{D(\beta\ell)} \quad , \quad B_4 = -\frac{1}{\beta}\frac{S_-(\beta\ell)}{D(\beta\ell)} \tag{2.167}$$

The shape function $N_4(x)$ then is

$$N_4(x) = \frac{1}{\beta\,D(\beta\ell)}\left[C_-(\beta\ell)\,S_-(\beta x) - S_-(\beta\ell)\,C_-(\beta x)\right] \tag{2.168}$$

This solution is similar to the expression (2.163) for $N_3(x)$.

The remaining shape functions $N_1(x)$ and $N_2(x)$ follow directly from symmetry considerations. In fact the shape function $N_1(x)$ is equal to $N_3(x')$,

and $N_2(x)$ is equal to $-N_4(x')$, where the coordinate $x' = \ell - x$ measures distance from the right end of the beam. Thus, these two shape functions are

$$N_1(x) = N_3(x') \quad , \quad N_2(x) = -N_4(x') \qquad (2.169)$$

This completes the determination of the shape functions for harmonic beam vibrations.

Box 2.17: Stiffness matrix for harmonic beam vibrations

The stiffness matrix K of a beam of length ℓ, bending stiffness EI, cross-section area A and mass density ρ, vibrating with angular frequency ω, is

$$\mathsf{K} = \frac{EI}{\ell^3} \begin{bmatrix} 12\psi_1 & -6\psi_3\ell & -12\psi_2 & -6\psi_4\ell \\ -6\psi_3\ell & 4\psi_5\ell^2 & 6\psi_4\ell & 2\psi_6\ell^2 \\ -12\psi_2 & 6\psi_4\ell & 12\psi_1 & 6\psi_3\ell \\ -6\psi_4\ell & 2\psi_6\ell^2 & 6\psi_3\ell & 4\psi_5\ell^2 \end{bmatrix}$$

The stiffness matrix depends on frequency via the wave number β,

$$\beta^4 = \omega^2 \frac{\rho A}{EI}$$

The coefficients $\psi_1, \cdots, \psi_6$ are functions of $\beta\ell$,

$$\psi_1 = \tfrac{1}{12}(\beta\ell)^2 \Psi \left[\cosh(\beta\ell)\sin(\beta\ell) + \sinh(\beta\ell)\cos(\beta\ell) \right]$$
$$\psi_2 = \tfrac{1}{12}(\beta\ell)^2 \Psi \left[\sinh(\beta\ell) + \sin(\beta\ell) \right]$$
$$\psi_3 = \tfrac{1}{6}(\beta\ell) \Psi \sinh(\beta\ell)\sin(\beta\ell)$$
$$\psi_4 = \tfrac{1}{6}(\beta\ell) \Psi \left[\cosh(\beta\ell) - \cos(\beta\ell) \right]$$
$$\psi_5 = \tfrac{1}{4} \Psi \left[\cosh(\beta\ell)\sin(\beta\ell) - \sinh(\beta\ell)\cos(\beta\ell) \right]$$
$$\psi_6 = \tfrac{1}{2} \Psi \left[\sinh(\beta\ell) - \sin(\beta\ell) \right]$$

with the common factor

$$\Psi = \frac{\beta\ell}{1 - \cosh(\beta\ell)\cos(\beta\ell)}$$

The columns of the stiffness matrix are the nodal force vectors corresponding to each of the mode shapes $N_1(x), \cdots, N_4(x)$. In beam bending theory the shear force and moment are determined by the third and second derivatives of the mode shape function, and thus the stiffness matrix can be expressed as

$$\mathsf{K} = EI \begin{bmatrix} N_1'''(0) & N_2'''(0) & N_3'''(0) & N_4'''(0) \\ N_1''(0) & N_2''(0) & N_3''(0) & N_4''(0) \\ -N_1'''(\ell) & -N_2'''(\ell) & -N_3'''(\ell) & -N_4'''(\ell) \\ -N_1''(\ell) & -N_2''(\ell) & -N_3''(\ell) & -N_4''(\ell) \end{bmatrix} \qquad (2.170)$$

The stiffness matrix now follows from differentiation of the shape functions given by (2.163), (2.168) and (2.169). The result is summarized in Box 2.17 in a format similar to that used in Box 2.15 for the beam on elastic springs.

All elements of the stiffness matrix are given as the value attained in the the static problem, multiplied by a coefficient $\psi_j(\beta\ell)$. Figure 2.65 shows these coefficients as function of $\beta\ell$. It is seen, that the coefficients ψ_1 and ψ_5, appearing on the diagonal, pass through zero and attain negative values. These particular values of $\beta\ell$ indicate free vibrations, in which the corresponding degree of freedom oscillates, while the other three are fixed. This is illustrated in the following example.

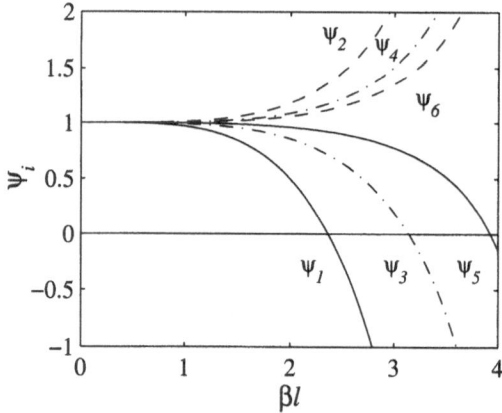

Figure 2.65: Coefficients $\psi_1, \cdots, \psi_6$ for harmonic beam vibrations.

Example 2.31

Figure 2.66 shows a beam acted upon by a moment M_B with harmonic time variation with angular frequency ω. The rotation θ_B and the mo-

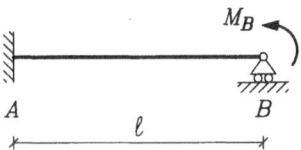

Figure 2.66: Beam with harmonic moment M_B.

ment M_B are related via the stiffness matrix element k_{44} as

$$M_B = k_{44}\,\theta_b$$

Thus, if $k_{44} = 0$ at some angular frequency ω_n, the rotation θ_B can exhibit harmonic variation, even at vanishing load, $M_B = 0$. This corresponds to a free vibration satisfying the homogeneous boundary conditions $w_A = w_B = 0$, $\theta_A = 0$ and $M_B = 0$.

The angular frequency of the free vibration is determined by the condition

$$k_{44} = \frac{4EI}{\ell}\,\psi_5(\beta\ell) = 0$$

According to Box 2.17 the equation for the free vibration wave number β_n is

$$\cosh(\beta_n\ell)\sin(\beta_n\ell) - \sinh(\beta_n\ell)\cos(\beta_n\ell) = 0$$

or by rearranging the terms

$$\tan(\beta_n\ell) = \tanh(\beta_n\ell)$$

This transcendental equation can be solved by plotting the graphs of the left and right sides and finding the abscissae of the points of intersection like in Example 2.30. The first roots are

$$\beta_n\ell = 3.927\ ,\ 7.069\ ,\ 10.210$$

and the higher roots are approximately determined from the equation $\tan(\beta_n\ell) \simeq 1$ as

$$\beta_n\ell \simeq \left(n + \tfrac{1}{4}\right)\pi \quad,\quad n = 4, 5, \cdots$$

The corresponding angular frequencies ω_n follow from (2.149).

2.10   Summary

The mechanics of beam bending is identified by considering the special case of a homogeneous beam, bent by opposing moments of equal magnitude at the ends. The beam bends into a circular shape, with an inner compression side and an outer extension side. A simple geometric argument establishes that the amount of extension or contraction, called the strain, is proportional to the distance from a neutral line, at which the length remains unchanged. If the beam is elastic, the stress is proportional to the strain, and integration of the stress over a cross-section establishes a linear relation between the moment and the curvature, as shown in Box 2.1.

The theory of elastic beams consists of the combination of static equilibrium equations and the moment-curvature relation. The general relations are given in Box 2.3, and the linearized equations for moderate rotations are given in Box 2.4.

If the internal forces in a beam can be determined from static equilibrium conditions alone, without reference to the deformation properties of the beam, the beam is called statically determinate. This property depends on the support conditions. In a statically determinate beam it is often convenient to determine the internal force distribution first from statics, and then to use this internal force distribution to obtain the beam displacement and rotation distribution. In this way the fourth order beam bending differential equation is replaced by a second order static equilibrium differential equation, and a second order differential equation for the displacement distribution. The two-step solution procedure is illustrated in Examples 2.3-2.5.

The beam bending problem possesses a special kind of similarity between the static equations for the internal force distributions and the corresponding set of kinematic equations for the displacement and rotation distributions. This similarity leads to the existence of a so-called virtual work equation, described in Section 2.4 and summarized in Box 2.5. Briefly speaking, virtual work is a statement about the work that would be performed by the external loads and the internal forces, if the beam were imagined to be displaced by infinitesimally small displacements. The virtual work equation says that the external work, performed by the actual loads through the virtual displacements, equals the internal work, performed by the internal forces through virtual deformation of the beam. The principle of virtual work plays a fundamental role in the modern theories of the mechanics solids and structures, and its importance as a property of the underlying mechanics principles can hardly be overemphasized. For this reason, the principle of virtual work, and a similar principle of complementary virtual work, in which the roles of an actual static field and a virtual kinematic field are interchanged, are derived in considerable generality in Section 2.4. In the present context the main application is to the calculation of the displacement at particular points by selection of a particular virtual static field, determined from a virtual load. The basic result is shown in Box 2.6, and the technique is illustrated in Examples 2.6-2.10.

If the internal force distribution depends on the deformation properties, the beam is called statically indeterminate. In principle this implies that analysis of statically indeterminate beams must use the full fourth order beam bending equation. In practice the full fourth order differential equation is rarely used. Instead the stiffness is calculated by use of the complementary virtual work technique. The distribution of the internal forces is then calculated by one of two techniques: either the structure is reduced to to an equivalent statically determinate structure with some additional local loads, or a set of deformations are imposed in such a way that they satisfy the conditions of the original structure. These two techniques – the force method and the deformation method – are summarized in Box 2.7 and Box 2.8, respectively. In the last decades interest has focused on a general formulation of the deformation method, called the Finite Element Method, in which a model of the structure is built from individual elements. In the present context the structure may be a frame, modeled by an assembly of individual beam elements. This technique is briefly described in Section 2.6 and the basic beam bending element is given in Box 2.9. A similar beam-column element, incorporating the effect of a normal force in the beam, is given in Box 3.3. A systematic extension of the classical beam bending theory to include shear flexibility is presented in Section 2.7, and the corresponding beam stiffness matrix is given in Box 2.12.

The remainder of the chapter treats two special problems, in which extra terms appear in the beam bending equation. The first class of problems is the so-called beam on elastic foundation, discussed in Section 2.8. The basic idea is that the beam rests on a medium that resists deformation. In the simplest form the resistance is equivalent to a distribution of springs, the so-called Winkler foundation. The extra term changes the properties of the solution, basically from a cubic polynomial to a set of decaying exponentials. This model gives a fairly accurate representation of the axisymmetric deformation patterns in cylindrical shells and tanks, illustrated in Example 2.25. In order to obtain a realistic representation of foundation problems it is necessary to include an extra term, accounting for the interaction between the motion of neighboring points. The calibration of such a representation to an elastic half space is discussed in some detail, and useful graphs provided.

The final section of the chapter deals with the dynamic behaviour of beams. The corresponding equation of motion is obtained by observing that the acceleration of a part of a beam imposes an inertial load on the beam, proportional to the mass per unit length and the acceleration. The basic equations are set out in Box 2.16, and the remainder of the section deals with harmonic motion of beams. A convenient representation of the solution is introduced and used to analyze the frequencies and mode shapes of free vibrations, and to derive the stiffness matrix of a beam element in harmonic motion.

2.11   Exercises

Exercise 2.1   Derive the moment distribution $M(x)$ from Example 1.3 by integration of the differential equation $\mathrm{d}^2M/\mathrm{d}x^2 = -p(x)$ with boundary conditions $M(0) = M(\ell) = 0$. Then find the shear force distribution $Q(x) = \mathrm{d}M/\mathrm{d}x$, and determine the reactions from $Q(0)$ and $Q(\ell)$.

Exercise 2.2   Determine and plot the moment distribution $M(x)$ in the Four-point beam bending shown in Fig. 2.7. Find the displacement $w(x)$ in the center span using the linearized curvature relation (2.10a), and verify the expression for the center displacement given in Example 2.2.

Exercise 2.3   The figure shows a beam of length ℓ and constant bending stiffness EI, fixed at A and simply supported at B. The beam is loaded by a concentrated moment M_B at B.

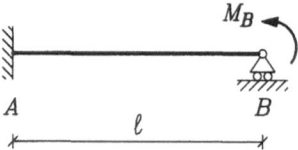

Find the displacement $w(x)$, the rotation $\theta(x)$, the moment $M(x)$, and the shear force $Q(x)$. Find in particular the rotation θ_B, and identify the rotation stiffness k_B at B, defined by $M_B = k_B \theta_B$. Compare the rotation stiffness k_B in the present case with that of a simply supported beam, determined in Example 2.3. (This solution is given in a slightly different form as A.4.2 in Appendix A)

Exercise 2.4   Consider a beam of length ℓ and constant bending stiffness EI with fixed ends, loaded by a distributed constant transverse force of intensity p. Derive the internal forces $M(x), Q(x)$, displacement $w(x)$ and rotation $\theta(x)$ by integration of the beam bending differential equations with appropriate boundary conditions. (Solution given as A.3.2 in Appendix A)

Exercise 2.5   Consider a simply supported beam of length ℓ and constant bending stiffness EI, loaded by a concentrated transverse force P at the distance a from the left end as shown in Fig. 2.13. Find the rotations θ_A and θ_B of the ends of the beam by the principle of complementary virtual work. Also use the principle of complementary virtual work to find the displacement $w(a)$ and the rotation $\theta(a)$ at the point of loading, $x = a$.

Exercise 2.6   Find the displacement function $w(x)$ by the principle of complementary virtual work for the simply supported beam with a concentrated transverse force P at distance a from the left end of the beam shown in Fig. 2.13. (The solution procedure was illustrated in Example 2.8. For $x \leq a$ the integral is evaluated from the integration formula of Appendix B by considering the three intervals $(0, x)$, (x, a), and (a, ℓ). The result for $x > a$ is obtained by interchanging x and a in the solution for $x \leq a$)

Exercise 2.7   Determine the vertical downward displacement u_C and the rotation θ_C of point C of the angle beam in Fig. 2.18. Increase the bending stiffness of the part AB from EI to $2EI$ and calculate the new values of the displacement w_C and θ_B dealt with in Example 2.10.

Exercise 2.8   Solve the problem of the uniformly loaded beam with three simple supports from Example 2.11 by introducing a hinge in the beam at the center C. The corresponding load X_1 is a moment couple, corresponding to the internal moment at the center of the beam. Determine the reactions

R_A, R_B, R_C, the internal moment distribution $M(x)$ and the displacement function $w(x)$. Use the sign of the moment X_1 to explain, why the center support carries more than half of the load.

Exercise 2.9   Consider a beam AB of length ℓ and constant bending stiffness EI with fixed ends, loaded by a distributed constant transverse force of intensity p. Determine the moments $M_A = M_B$ at the ends of the beam by using the end rotations of a similar simply supported beam, determined in Example 2.4, in the symmetric rotation spring solution from Example 2.12. Use these end moments to determine the internal moment distribution $M(x)$ and then the shear force distribution $Q(x)$. (The solution is given as A.3.2 in Appendix A)

Exercise 2.10   Use the force method or a special solution from Example 2.12 to determine the moment over the center support and the three reactions in the two-span beam shown in the figure.

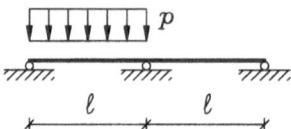

Exercise 2.11   Use the force method to determine the moment over the center supports and the four reactions for the two load cases of the three-span beam shown in the figure.

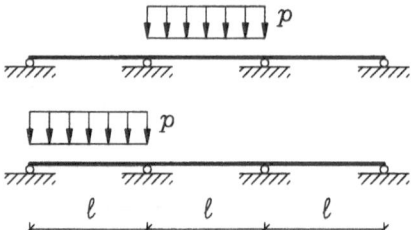

Exercise 2.12   Plot the internal moment $M(x)$ for the first few spans of the semi-infinite beam shown in Fig. 2.23 and treated in Example 2.13. Use the differential relation $Q = \mathrm{d}M/\mathrm{d}x$ to determine the (constant) shear force Q_j in each of the spans. Find the vertical reactions R_j from each of the

supports $j = 0, 1, 2, \ldots$, and show by summation that the sum of all the reactions is zero.

Exercise 2.13 Consider one or two uniformly loaded spans in an infinite beam on equally spaced simple supports, shown in the figure. Determine the moments over the supports of the loaded spans in both cases, and find the reactions through these supports. What changes would happen, if the connections with the infinite unloaded parts of the beam were cut.

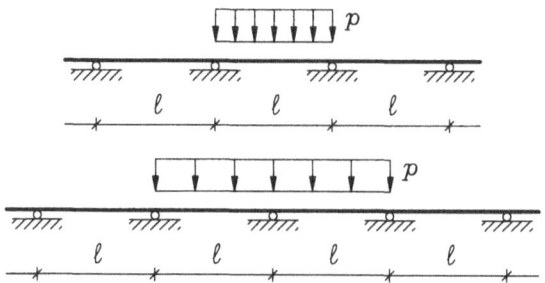

Exercise 2.14 Complete the analysis of the beam from Example 2.16, by determining the internal moment distribution and the reactions for the special case of equal spans, $a = b$.

Exercise 2.15 Find the displacement at the center of a simply supported beam of length ℓ, bending stiffness EI and shear stiffness GA_0 when the beam is loaded by a concentrated transverse force P at the center. (See Box 2.11 and Example 2.22.)

Exercise 2.16 The figure in Example 2.31 shows a beam with one end fixed and the other rotated by an external moment M_B. The stiffness giving the moment M_B as function of the rotation θ_B is one of the elements in the beam element stiffness matrix - which one? Give the rotation stiffness for a beam element with and without shear flexibility. The shear flexibility is represented via the non-dimensional parameter $\Phi = 12EI/GA_0\ell^2$. Find the magnitude of the parameter Φ that corresponds to a 5 pct. reduction of the rotation stiffness.

Exercise 2.17 Consider an infinitely long beam with bending stiffness EI supported by distributed springs of stiffness k. According to Example 2.23 the displacement $w(x)$ from a concentrated force P located at t is $w(x) = \frac{1}{2}\lambda(P/k)A(\lambda(x - t))$, where the function $A(\ )$ is defined in Box 2.13. Let the beam be loaded by a constant distributed load $p(t) = p_0$ in the interval $[-a, a]$. Find the displacement $w_a(0)$ and the moment $M_a(0)$ at the center of the loaded interval in terms of the total load $P = 2ap_0$. The corresponding displacement $w_0(0)$ and moment $M_0(0)$ under a concentrated force P were determined in Example 2.23. Find the reduction factors $w_a(0)/w_0(0)$ and $M_a(0)/M_0(0)$ due to the spread of the load. (Use the differentiation/integration relations in Box 2.13 to obtain $w_a(0)$ and $M_a(0)$.)

Exercise 2.18 Consider a beam of length ℓ with bending stiffness EI and distributed spring support with stiffness k. Let the beam be loaded by a constant transverse load $p(x) = p_0$. Set up the appropriate differential equation and boundary conditions for the case of free ends. Introduce a coordinate system with origo at the center of the beam and solve the differential equation by combining a particular solution for the constant load with a symmetric solution to the homogeneous equation. (Suitable symmetric solutions to the homogeneous equation are $\cosh(\lambda x)\cos(\lambda x)$ and $\sinh(\lambda x)\sin(\lambda x)$.)

Exercise 2.19 Consider a cylindrical fluid filled tank of height h similar to the one shown in Fig. 2.47 of Example 2.25, but with fully constrained support at ground level, $w(0) = 0$ and $\theta(0) = 0$. The particular solution $w_\gamma(x)$ for fluid load, but disregarding the boundary conditions, is given in Example 2.25. Impose the boundary constraints, and obtain the solution $w(x), T(x), M(x)$ in terms of the functions $A(\lambda x), \cdots, D(\lambda x)$ from Box 2.13. Plot or sketch the hoop force $T(x)$ and the moment $M(x)$ for $\lambda h = 5$ as in Fig. 2.47.

Exercise 2.20 Consider a rigid beam of length ℓ resting on a composite elastic foundation with parameters k and g. Let the beam be loaded by a uniform transverse load of magnitude p_0. Use Fig. 2.53 and the formula (2.131) for P_0 to find the fraction of the total load $p_0\ell$ that is carried by the part of the foundation outside the beam. Determine the uniform displacement w_0 of the beam.

Exercise 2.21 Use the technique of Example 2.27 to obtain the solution to the problem of a homogeneous beam of length ℓ resting on a composite elastic foundation with parameters k and g, when the load $p(x)$ on the beam increases linearly from $-p_0$ at the left end and p_0 at the right end of the beam,

$$p(x) = p_0 \frac{x}{\ell} - p_0 \frac{x'}{\ell}$$

(The moment distribution, and thereby the curvature $\mathrm{d}^2 w/\mathrm{d}x^2$, is represented by the anti-symmetric function $[\sinh(\lambda_+ x')\sin(\lambda_- x) - \sinh(\lambda_+ x)\sin(\lambda_- x')]$)

Exercise 2.22 Consider the problem of harmonic free vibrations of a homogeneous beam of length ℓ with bending stiffness EI, cross-section area A and mass density ρ. Let the boundary conditions be

$$w_A = 0 \quad , \quad w'_A = 0$$

at the left end, and

$$w_B = 0 \quad , \quad -\frac{M_B}{EI} = w''_B = 0$$

at the right end. Represent the solution in terms of the functions $S_+(\beta x), \cdots,$ $C_-(\beta x)$, defined in (2.151). Obtain the equation for free vibrations in terms of $\beta_n \ell$,

$$\tan(\beta_n \ell) = \tanh(\beta_n \ell)$$

Illustrate this equation in a graphical format similar to that used in Example 2.30, and obtain the first root $\beta_n \ell$. Find an expression for displacement function $w_n(x)$, giving the shape of free vibrations corresponding to the wave number β_n. (This problem was treated in an alternative way in Example 2.31)

Exercise 2.23 Consider the problem of harmonic free vibrations of a homogeneous beam of length ℓ with bending stiffness EI, cross-section area A and mass density ρ. Let the boundary conditions be

$$w_A = 0 \quad , \quad w'_A = 0$$

at the left end, and

$$-\frac{Q_B}{EI} = w'''_B = 0 \quad , \quad w'_B = 0$$

at the right end. Represent the solution in terms of the functions $S_+(\beta x), \cdots,$ $C_-(\beta x)$, defined in (2.151). Obtain the equation for free vibrations in terms of $\beta_n \ell$. Illustrate the equation in a graphical format similar to that used in Example 2.30, and obtain the first root $\beta_n \ell$. Give an approximate formula for the higher roots $\beta_n \ell$, $n > 3$. Find an expression for displacement function $w_n(x)$, giving the shape of free vibrations corresponding to the wave number β_n. (The first root $\beta_1 \ell = 2.365$ corresponds $\psi_1 = 0$, shown in Fig. 2.65)

Chapter 3

Columns

A slender structural member with a substantial axial load component is usually called a column. In the beam theories treated in Chapter 2 the internal forces could be determined to a good approximation without accounting for the deformed state of the beam. This is different in the case of a column, where a transverse displacement changes the position of the cross-section with respect to the normal force, and thereby introduces an additional moment. This may lead to instability of the column, when the normal force becomes sufficiently large in compression. The basic problem of a curve with elastic bending stiffness - the so called 'elastica' - loaded by a concentrated force at the end was treated by LEONHARD EULER (1644), who gave a very extensive and general analysis of this special problem.

Often the normal force is an effect in addition to other effects like transverse load or bending by end moments. Slender structural members, where the normal force appears together with other effects, are called beam-columns. In this chapter the theory of elastic columns and beam-columns is treated in four steps. First the simple bending theory of beams is extended to include the effect of a normal force in Section 3.1. It is demonstrated that the application of a compressive normal force leads to reduced stiffness of the beam-column. For a sufficiently large compressive load there is no stiffness to restrain bending, and the column becomes unstable and buckles.

The classical column problem, consisting of determination of critical loads, at which instability occurs, and determination of the associated buckling shapes, is treated in Section 3.2. In practice columns are not ideally straight, and the load is not only axial. Buckling of real columns is therefore a gradual process, in which the displacements increase, and eventually become virtually unbounded at the critical loads, treated in the idealized column problem. The magnitude of the displacements before reaching the critical load is determined by imperfections in the initial geometry and bending loads, causing initial curvature. The role of imperfections is discussed in Section 3.3, and the load carrying capacity of an imperfect column is determined.

The beam-column bending problem is treated in Section 3.4, where the stiffness matrix of a beam-column is derived, and its use in the solution of some simple problems is illustrated.

The theory presented in Sections 3.1 to 3.4 is essentially linear, if the normal force is assumed known. This theory forms the basis of most column and beam-column calculations used in practice. However, the fact that actual displacements are only accounted for in an approximate manner, limits the ability of the theory to treat the behaviour at the critical load, where displacements become large. Therefore a brief account of the general 'elastica' problem is given in Section 3.5 in order to identify basic features of the complete solution, e.g. the ability to carry load after buckling.

3.1   Beam with normal force

The basis for the theory of elastic beam-columns is the equilibrium equations, formulated for the deformed state of the beam. Figure 3.1 shows the deformed state of a beam column with distributed transverse load $p(x)$. The displacements, and in particular the rotations due to the displacement gradients, are assumed to be small, and thus no distinction will be made between the length increment ds along the beam axis in the deformed state and its projection dx on the line of the initial beam axis. The figure shows a slice of thickness $ds \simeq dx$, and the forces and moments acting on it. The internal force vector on a section of the beam has the components (N, Q) in the axial and transverse directions, respectively. Note, that in the present formulation of a linearized beam-column theory the normal force N is taken as the component along the direction of the original beam axis, and the shear

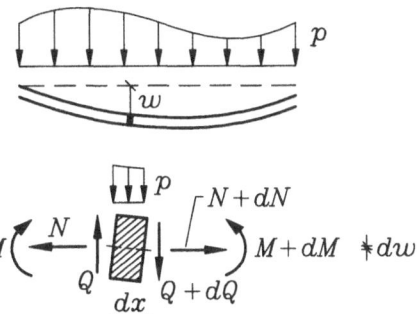

Figure 3.1: Equilibrium of thin beam section in displaced position.

force Q is in a direction normal to this. In the linearized theory discussed
in Sections 3.1–3.4, the normal force will be treated as prescribed, and the
axial equilibrium and extension of the beam will not be treated separately.

Equilibrium in the transverse direction requires that the projection of all
forces on this direction have the sum zero. With the present definition of
shear and normal force only the shear force Q contributes to the transverse
equilibrium, where the sum of internal and external contributions is

$$(Q + dQ) - Q + p\,dx = 0 \tag{3.1}$$

The terms $\pm Q$ cancel, and division by dx leads to the differential equation

$$\frac{dQ}{dx} = -p \tag{3.2}$$

The sum of moments must also vanish. In the present case the shear force
Q contributes as a force couple with distance dx, and the normal force N
contributes as a force couple with distance dw. This gives the moment equilibrium equation

$$(M + dM) - M + N\,dw - Q\,dx = 0 \tag{3.3}$$

Cancellation of $\pm M$ and division by dx leads to the differential equation

$$\frac{dM}{dx} + N\frac{dw}{dx} = Q \tag{3.4}$$

The two first order differential equations (3.2) and (3.4) must be satisfied
irrespective of the material properties of the beam. The shear force can be

eliminated, resulting in the second order differential equation

$$\frac{d^2 M}{dx^2} + \frac{d}{dx}\left(N \frac{dw}{dx}\right) + p = 0 \tag{3.5}$$

This differential equation contains the moment M, the normal force N and the displacement derivative dw/dx. The dependence on the displacement gradient excludes the possibility of determining the internal forces independent of the displacements. It is therefore necessary to express the moment M in terms of the deformation of the beam.

In the linearized beam bending theory of Sections 2.1–2.2 the cross-section rotation θ and the curvature of the beam axis κ were introduced as

$$\theta = -\frac{dw}{dx} \quad , \quad \kappa = \frac{d\theta}{dx} = -\frac{d^2 w}{dx^2} \tag{3.6}$$

The relation between the moment M and the curvature κ is not changed by the presence of the normal force, and thus the moment is expressed in terms of the bending stiffness EI and the curvature by (2.19),

$$M = EI \kappa = -EI \frac{d^2 w}{dx^2} \tag{3.7}$$

When this expression is substituted into (3.5), the following fourth order equation is obtained for the elastic beam-column.

$$\frac{d^2}{dx^2}\left(EI \frac{d^2 w}{dx^2}\right) - \frac{d}{dx}\left(N \frac{dw}{dx}\right) - p = 0 \tag{3.8}$$

In the following the normal force N will be considered as a parameter or known function, and thus the beam-column differential equation (3.8) is linear. The general non-linear 'elastica' is treated in Section 3.5.

The beam-column equation (3.8) must be solved in connection with two boundary conditions at each end. These boundary conditions can be either kinematic, when expressed in terms of the displacement w and the cross-section rotation θ, or static, when expressed in terms of the moment M and the shear force Q. In the case of static boundary conditions the moment and shear force are expressed in terms of the displacement by use of the relations (3.7) and (3.4), respectively. The differential equation and boundary conditions of the beam-column problem are summarized in Box 3.1.

The beam-column equation is used in two different contexts: either as an extended beam equation, in which the presence of the normal force modifies

---

Box 3.1: Differential equation of beam-column bending

The linearized differential equation for bending of an elastic beam-column is obtained by introducing linear approximations for the cross-section rotation θ and beam axis curvature κ,

$$\theta = -\frac{\mathrm{d}w}{\mathrm{d}x} \quad , \quad \kappa = \frac{\mathrm{d}\theta}{\mathrm{d}x} = -\frac{\mathrm{d}^2 w}{\mathrm{d}x^2}$$

and substituting the elastic beam bending relation

$$M = -EI\frac{\mathrm{d}^2 w}{\mathrm{d}x^2}$$

into the equilibrium equation for the deformed geometry of the beam-column.

This gives the 4'th order elastic beam-column differential equation

$$\frac{\mathrm{d}^2}{\mathrm{d}x^2}\left(EI\frac{\mathrm{d}^2 w}{\mathrm{d}x^2}\right) - \frac{\mathrm{d}}{\mathrm{d}x}\left(N\frac{\mathrm{d}w}{\mathrm{d}x}\right) - p = 0$$

to be solved with kinematic boundary conditions in terms of w and $\mathrm{d}w/\mathrm{d}x$, and static boundary conditions in terms of

$$M = -EI\frac{\mathrm{d}^2 w}{\mathrm{d}x^2} \quad , \quad Q = -\frac{\mathrm{d}}{\mathrm{d}x}\left(EI\frac{\mathrm{d}^2 w}{\mathrm{d}x^2}\right) + N\frac{\mathrm{d}w}{\mathrm{d}x}$$

For $p(x) = 0$ and homogeneous boundary conditions the beam-column equation may determine critical values N_c corresponding to instability. These critical values correspond to compression, i.e. $N_c < 0$.

---

the solution of the ordinary beam bending equation, or in the solution of a homogeneous column problem without bending, in which the normal force is treated as an unknown parameter that can reach a critical value, where buckling of the column occurs. This double role of the normal force is similar to the role of inertial forces, treated in Section 2.9, where the equation (2.147) can be used to deal with harmonic forced vibrations or to determine the frequencies and mode shapes of free vibrations. The role of the normal force as a modification of the beam bending problem is illustrated in the following example.

Example 3.1

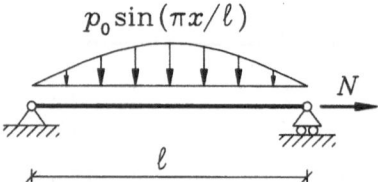

$$p_0 \sin(\pi x/\ell)$$

$$N$$

$$\ell$$

Figure 3.2: Beam-column with sine-distributed transverse load.

Figure 3.2 shows a simply supported beam-column with normal force N, loaded by a distributed transverse load of intensity

$$p(x) = p_0 \sin\left(\pi\frac{x}{\ell}\right) \quad , \quad 0 \le x \le \ell$$

The simple supports at the ends imply that $w = 0$ and $w'' = 0$ at $x = 0$ and $x = \ell$. These boundary conditions are satisfied by the function

$$w(x) = w_c \sin\left(\pi\frac{x}{\ell}\right) \quad , \quad 0 \le x \le \ell$$

and the displacement w_c at the center is determined by substitution into the differential equation (3.8),

$$EI\left(\frac{\pi}{\ell}\right)^4 w_c + N\left(\frac{\pi}{\ell}\right)^2 w_c - p_0 = 0$$

This determines the center displacement as

$$w_c = \frac{p_0}{EI\left(\frac{\pi}{\ell}\right)^4 + N\left(\frac{\pi}{\ell}\right)^2}$$

The result may be written in the alternative form

$$w_c = \frac{1}{1 + \dfrac{N}{EI}\left(\frac{\ell}{\pi}\right)^2} \frac{p_0}{EI}\left(\frac{\ell}{\pi}\right)^4$$

The last factors represent the displacement in the corresponding beam problem with $N = 0$,

$$w_c^0 = \frac{p_0}{EI}\left(\frac{\ell}{\pi}\right)^4$$

The first factor is an amplification factor, containing the effect of the normal force. It is seen that the amplification becomes infinite at a compressive force of magnitude

$$P_E = EI \left(\frac{\pi}{\ell}\right)^2$$

This particular load is called the *Euler load*, a reference to the original work of Euler on columns. In column problems it is often convenient to consider compressive axial forces as positive. This is handled by introducing the notation $P = -N$, whereby P denotes an axial force with positive values corresponding to compression.

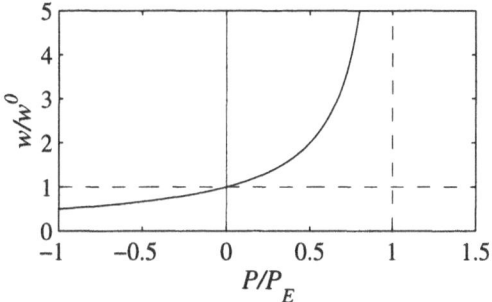

Figure 3.3: Beam-column amplification relation.

In terms of the two characteristic parameters w_c^0 and P_E the center displacement formula takes the form

$$w_c = \frac{w_c^0}{1 - \dfrac{P}{P_E}}$$

This relation is shown graphically in Fig. 3.3. Although the solution is particularly simple in the present case due to the special choice of load distribution, the amplification behaviour illustrated in Fig. 3.3 applies to most beam column-problems. It is seen that the application of axial compression $(P > 0)$ leads to an increase of the deformations, while axial tension $(P < 0)$ reduces the deformations. The effect of axial compression is much more dramatic than axial tension, and for $P = P_E$ the beam-column has lost its stiffness completely, leading to column instability, discussed in Section 3.2.

3.2   Ideal straight columns

In Example 3.1 it was found that for a sufficiently large compressive axial
force a simply supported beam could obtain arbitrarily large transverse deformations, even for a very small transverse load. This axial load, often called
the Euler load, can be identified directly, without applying a transverse load.
The idea is to consider an ideally straight column as shown in Fig. 3.4a. A
compressive axial load $P = -N > 0$ is then applied, Fig. 3.4b. Hereby the
column becomes slightly shorter, but in most cases of practical interest, this
shortening is unimportant. The main point is, that because the column is
ideally straight and there is no transverse load, it will remain straight under
a limited axial load. If the axial load is increased, a magnitude P_E is reached,
at which two solutions exist: a straight configuration, and a buckled form as
shown in Fig. 3.4c. This problem has the form of an *eigenvalue problem*, and
the associated critical load P_E is found as an eigenvalue.

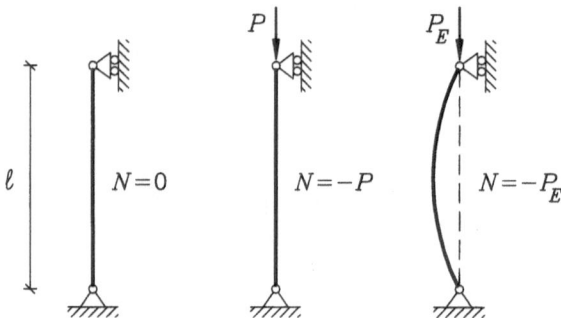

Figure 3.4: The Euler column.

Consider a column of length ℓ and constant bending stiffness EI. There is
no transverse load, and the differential equation (3.8) is conveniently written
in the normalized form

$$\frac{d^4w}{dx^4} + k^2 \frac{d^2w}{dx^2} = 0 \tag{3.9}$$

where the parameter k is introduced via

$$k^2 = \frac{P}{EI} \tag{3.10}$$

The general solution to this homogeneous 4'th order differential equation is

$$w(x) = C_1 + C_2\,kx + C_3\,\cos(kx) + C_4\,\sin(kx) \qquad (3.11)$$

In this form the coordinate x only appears in the non-dimensional combination kx. The moment and shear force follows from Box 3.1 as

$$\frac{M(x)}{EI} = -\frac{\mathrm{d}^2 w}{\mathrm{d}x^2} = C_3\,k^2 \cos(kx) + C_4\,k^2 \sin(kx) \qquad (3.12)$$

and

$$\frac{Q(x)}{EI} = -\frac{\mathrm{d}^3 w}{\mathrm{d}x^3} - k^2\frac{\mathrm{d}w}{\mathrm{d}x} = -C_2\,k^3 \qquad (3.13)$$

These relations are used to formulate static boundary conditions.

The boundary conditions of the Euler column shown in Fig. 3.4 are

$$w(0) = w(\ell) = 0 \quad , \qquad M(0) = M(\ell) = 0 \qquad (3.14)$$

Note, that the differential equation (3.9) and the boundary conditions (3.14) are homogeneous. Thus, the solution will be $w(x) \equiv 0$, except for particular values of the parameter k that permit a nontrivial solution. These values k_n are the eigenvalues, and to each eigenvalue corresponds an eigenfunction $w_n(x)$. The eigenfunctions describe the mode of buckling deformation, and are therefore also called buckling modes.

The boundary conditions at $x = 0$ give the equations

$$\begin{aligned} w(0) &= C_1 + C_3 = 0 \\ w''(0) &= k^2 C_3 = 0 \end{aligned} \qquad (3.15)$$

from which $C_1 = C_3 = 0$. The boundary conditions at $x = \ell$ then give

$$\begin{aligned} w(\ell) &= k\ell\,C_2 + \sin(k\ell)\,C_4 = 0 \\ w''(\ell) &= -k^2 \sin(k\ell)\,C_4 = 0 \end{aligned} \qquad (3.16)$$

These equations imply, that

$$k\ell\,C_2 = 0 \quad , \qquad k^2 \sin(k\ell)\,C_4 = 0 \qquad (3.17)$$

The representation of the general solution in the form (3.11) is based on the assumption that $P > 0$, and thereby $k\ell > 0$. It then follows from (3.17a), that $C_2 = 0$. This leaves the final equation (3.17b). Naturally this equation can be satisfied by $C_4 = 0$, but this would reduce the solution to $w(x) \equiv 0$.

A nontrivial solution with $C_4 \neq 0$ is found by selecting the parameter k such that

$$\sin(k\ell) \; = \; 0 \qquad\qquad (3.18)$$

This equation has the positive roots

$$k\ell \; = \; \pi, 2\pi, 3\pi, \cdots \quad\quad \text{or} \quad\quad k_n \; = \; n\frac{\pi}{\ell} \; , \; n = 1, 2, 3, \cdots \qquad (3.19)$$

These roots correspond to the axial loads

$$P_n \; = \; EI\,k_n^2 \; = \; n^2 \Big(\frac{\pi}{\ell}\Big)^2 EI \quad , \quad n = 1, 2, 3, \cdots \qquad (3.20)$$

The smallest of these loads is called the Euler load,

$$P_E \; = \; \Big(\frac{\pi}{\ell}\Big)^2 EI \qquad\qquad (3.21)$$

At this load the column can buckle into a non-straight mode of deformation, given by the transverse displacement

$$w_E(x) \; = \; C \sin\Big(\pi\frac{x}{\ell}\Big) \qquad\qquad (3.22)$$

Note, that the transverse displacement of a beam-column with a transverse load will grow towards infinity, as the axial compression force P approaches the Euler load P_E as demonstrated for a special case in Example 3.1. Thus, it appears that a beam column gradually looses its bending stiffness with increasing normal compression.

In general the buckling modes of the Euler column and the corresponding buckling loads are given by

$$w_n(x) \; = \; C_n \sin\Big(n\pi\frac{x}{\ell}\Big) \quad , \quad P_n \; = \; n^2 \Big(\frac{\pi}{\ell}\Big)^2 EI \quad , \quad n = 1, 2, 3, \cdots \quad (3.23)$$

The first three buckling modes are shown in Fig. 3.5. In practice it will be difficult to increase the load beyond the smallest buckling load P_E, if the column is only supported at the ends. However, the higher buckling modes correspond to the buckling modes of columns with equally spaced intermediate supports.

The column theory expressed by the linear differential equation (3.9) is only approximate. In its derivation it was assumed that the rotations are 'small', and that the length along the deformed beam can be represented by its projection, $ds \simeq dx$. These approximations reduce the problem to the form

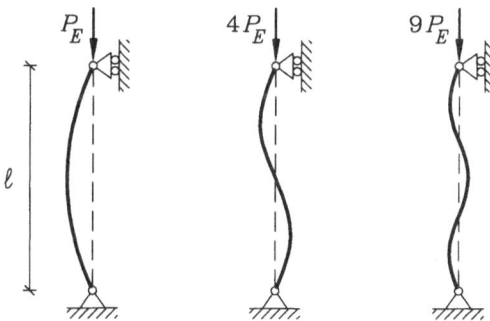

Figure 3.5: First three buckling modes of the Euler column.

of a linear eigenvalue problem, but also limit the scope of the solution to the onset of instability, where the deformation is small. Thus, the theory is useful in establishing a reference value, such as P_E, for the onset of instability, while description of the development of the load and displacements after the onset of instability requires a non-linear theory. A fully non-linear theory of the Euler column, the so-called elastica, is presented in Section 3.4.

Example 3.2

Figure 3.6 shows a column of length ℓ with one fixed end, supporting an axial compression force P at the free end. The general solution is given by (3.11), where the arbitrary constants and the parameter k are to be determined by the boundary conditions as in the case of the simply

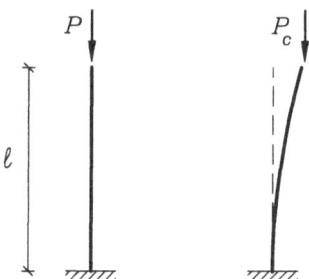

Figure 3.6: Column with one fixed end.

supported column treated above. In the present problem the boundary conditions at the fixed end are

$$w(0) \ = \ C_1 \ + \ C_3 \ = \ 0$$
$$w'(0) \ = \ k\,C_2 \ + \ k\,C_4 \ = \ 0$$

These equations give $C_3 = -C_1$, $C_4 = -C_2$, and the solution reduces to the form

$$w(x) \ = \ C_1[\,1 - \cos(kx)\,] \ + \ C_2[\,kx - \sin(kx)\,]$$

The boundary conditions at the top of the column are $M(\ell) = 0$ and $Q(\ell) = 0$. By (3.13) the conditions $Q(\ell) = 0$ gives $C_2 = 0$, and by (3.12) the condition $M(\ell) = 0$ then is

$$w''(\ell) \ = \ k^2 \cos(k\ell)\,C_1 \ = \ 0$$

From this equation the eigenvalues are determined as

$$k\ell \ = \ \tfrac{1}{2}\pi, \tfrac{3}{2}\pi, \tfrac{5}{2}\pi, \cdots \qquad \text{or} \qquad k_n \ = \ n\Big(\frac{\pi}{2\ell}\Big) , \ n = 1, 3, 5, \cdots$$

corresponding to the axial loads

$$P_n \ = \ EI\,k_n^2 \ = \ n^2\Big(\frac{\pi}{2\ell}\Big)^2 EI \ , \quad n = 1, 3, 5, \cdots$$

and the buckling modes

$$w_n(x) \ = \ C_n[\,1 - \cos(k_n x)\,] \ , \quad k_n \ = \ n\Big(\frac{\pi}{2\ell}\Big) , \ n = 1, 3, 5, \cdots$$

Note, that these modes correspond to the symmetric buckling modes of a simply supported Euler column of length 2ℓ, obtained by extending the actual column symmetrically below the fixed support.

Example 3.3

In Fig 3.7 a simple support has been added to the column of Example 3.2. This does not change the solution procedure, but the result can no longer be given explicitly. The general solution is given by (3.11), and after imposing the boundary conditions at the fixed end as in Example 3.2 the solution takes the form

$$w(x) \ = \ C_1[\,1 - \cos(kx)\,] \ + \ C_2[\,kx - \sin(kx)\,]$$

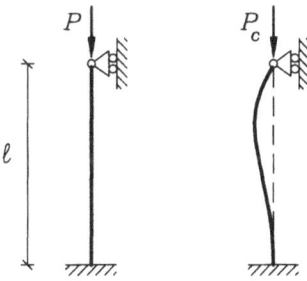

Figure 3.7: Column with fixed and simply supported ends.

The boundary conditions at the top of the column are $w(\ell) = 0$ and $M(\ell) = 0$, whereby

$$w(\ell) = [1 - \cos(k\ell)]\,C_1 + [k\ell - \sin(k\ell)]\,C_2 = 0$$
$$w''(\ell) = k^2 \cos(k\ell)\,C_1 + k^2 \sin(k\ell)\,C_2 = 0$$

A nontrivial solution to this pair of equations can only be obtained, if the determinant of the equation system vanishes, i.e. if

$$[1 - \cos(k\ell)]\,\sin(k\ell) - [k\ell - \sin(k\ell)]\,\cos(k\ell) = 0$$

This equation can be reformulated as

$$\tan(k\ell) = k\ell$$

This is a transcendental equation like the one treated in Example 2.30. The left and right hand sides are shown in Fig. 3.8.

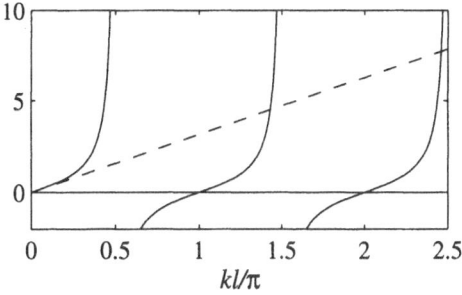

Figure 3.8: Stability equation: $\tan(k\ell)$ (–) and $k\ell$ (- -).

Box 3.2: Buckling load and effective length of simple columns

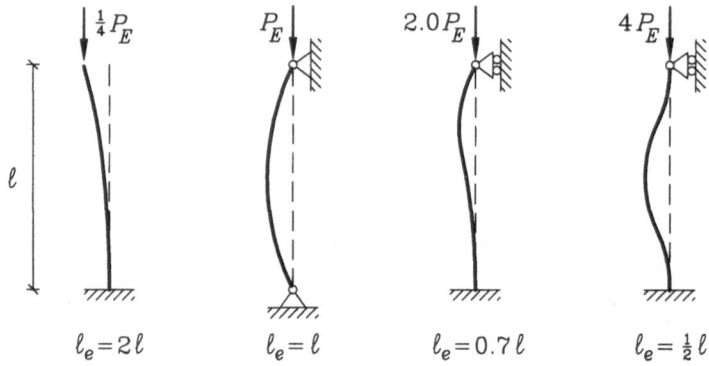

The critical load of an ideal simply supported column is given by the Euler load

$$P_E = EI\left(\frac{\pi}{\ell}\right)^2$$

The critical load P_c of ideal columns with different support conditions can be expressed in a similar format in terms of an *effective length* ℓ_e, defined such that

$$P_c = EI\left(\frac{\pi}{\ell_e}\right)^2 \quad \text{or} \quad \frac{P_c}{P_E} = \left(\frac{\ell}{\ell_e}\right)^2$$

The effective length ℓ_e is the distance between the inflection points, i.e. the points of change of curvature, of the buckling mode. It can often be estimated approximately by inspection.

The roots of the transcendental stability equation are given by the abscissae $k_1\ell, k_2\ell, \cdots$ of the points of intersection. These abscissae can be found by iteration, starting from the values at which $\tan(k\ell)$ has a vertical asymptote,

$$k_n^{(0)}\ell = \left(n + \tfrac{1}{2}\right)\pi \quad , \qquad n = 1, 2, 3, \cdots$$

Note, that the root $k_0\ell = 0$, corresponding to $k_0^0 = \tfrac{1}{2}\pi$ is without interest. The iteration procedure can be formulated as

$$k_n^{(i+1)}\ell = n\pi + \tan^{-1}(k_n^i\ell)$$

The two first steps and the final value of the parameter $k_n \ell$ are given below, together with the buckling loads, determined from (3.10).

n	1	2	3	4
$k_n^{(0)}\ell$	4.7124	7.8540	10.9956	14.1372
$k_n^{(1)}\ell$	4.5033	7.7273	10.9049	14.0665
$k_n \ell$	4.4934	7.7253	10.9041	14.0662
P_n/P_E	2.0457	6.0468	12.0471	20.0472

The table illustrates that, apart from the first two roots, the remaining roots are given to within 1 pct. by the formula $k_n \ell \simeq (n + \frac{1}{2})$, $n = 3, 4, \cdots$, used as the start value in the iteration. However, in a technical context the first root is of most interest.

3.3   Imperfect columns

In real life columns are not ideally straight. Figure 3.9a shows a column where the (small) deviation from a straight line in the unloaded state is described by the function $w_0(x)$. The length of the column is ℓ and the constant bending stiffness EI. When loaded by an axial compression force P, as shown in Fig. 3.9b an *additional* transverse displacement $w(x)$ occurs. The moment is determined by this additional displacement as

$$M = -EI \frac{\mathrm{d}^2 w}{\mathrm{d}x^2} \tag{3.24}$$

The effect of the normal force on the equilibrium is through the *total* displacement $w(x)+w_0(x)$, and thus the equilibrium equation takes the form

$$\frac{\mathrm{d}^2}{\mathrm{d}x^2}\left(EI \frac{\mathrm{d}^2 w}{\mathrm{d}x^2}\right) - \frac{\mathrm{d}}{\mathrm{d}x}\left(N \frac{\mathrm{d}w + \mathrm{d}w_0}{\mathrm{d}x}\right) = 0 \tag{3.25}$$

For constant axial force and bending stiffness this equation can be written as

$$\frac{\mathrm{d}^4 w}{\mathrm{d}x^4} + k^2 \frac{\mathrm{d}^2 w}{\mathrm{d}x^2} = -k^2 \frac{\mathrm{d}^2 w_0}{\mathrm{d}x^2} \tag{3.26}$$

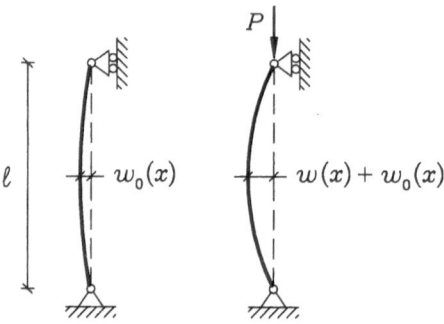

Figure 3.9: The Euler column with initial imperfection $w_0(x)$.

where the parameter $k^2 = P/EI$ was introduced in (3.10). The boundary conditions are $w(0) = w(\ell) = 0$ and $M(0) = M(\ell) = 0$.

It is seen that the initial imperfection $w_0(x)$ introduces a non-homogeneous term in the column equation, and thus the imperfect column problem is not an eigenvalue problem. The complete solution to the non-homogeneous differential equation (3.26) is obtained by representing $w(x)$ as a series expansion in terms of the eigenfunctions of the corresponding homogeneous equation. For the present problem the corresponding eigenvalue problem was solved in Section 3.2 and the solution given in (3.23). The series representation of the displacement $w(x)$ then takes the form

$$w(x) = \sum_{n=1}^{\infty} w_n \sin(k_n x) \quad , \quad k_n = n\frac{\pi}{\ell} \qquad (3.27)$$

Substitution of this expansion into the non-homogeneous equilibrium equation (3.26) gives

$$\sum_{n=1}^{\infty} (k_n^4 - k^2 k_n^2) w_n \sin(k_n x) = -k^2 \frac{d^2 w_0}{dx^2} \qquad (3.28)$$

The unknown displacement coefficients w_n are determined by representing the initial imperfection $w_0(x)$ as a series similar to the expansion (3.27) for $w(x)$.

$$w_0(x) = \sum_{n=1}^{\infty} w_n^0 \sin(k_n x) \quad , \quad k_n = n\frac{\pi}{\ell} \qquad (3.29)$$

The coefficients w_n^0 in this expansion are determined by use of the orthogonality relation for the sine function as

$$w_n^0 = \frac{2}{\ell} \int_0^\ell w_0(x) \sin(k_n x) \, dx \quad , \quad n = 1, 2, \cdots \qquad (3.30)$$

When the series expansion (3.29) for $w_0(x)$ is substituted into (3.28), the displacement coefficients are found to be

$$w_n = \frac{w_n^0}{\dfrac{k_n^2}{k^2} - 1} = \frac{w_n^0}{\dfrac{P_n}{P} - 1} \quad , \quad n = 1, 2, \cdots \qquad (3.31)$$

This result shows that for increasing tension, $P \to -\infty$, the column becomes increasingly straight, $w_n \to 0$. If a compression force $P > 0$ approaches any of the buckling loads P_n the corresponding component of the initial imperfection is amplified. In principle infinite amplification is obtained, when $P = P_n$.

The present analysis was applied to the special case of a simply supported column. However, the method of expanding both the unknown displacement function $w(x)$ and the initial imperfection function $w_0(x)$ in terms of the buckling modes of the corresponding homogeneous problem corresponding to an ideal straight column is also valid for other boundary conditions, and in each case the buckling modes also satisfy a suitable orthogonality relation. This leads to the general conclusion, that the application of an axial compression load P on an imperfect column will lead to amplification of the contributions of the different buckling mode components in the initial imperfection function $w_0(x)$. In practice the imperfection component corresponding to the first buckling mode will most often dominate the deformation. This leads to a column design procedure described in the following example.

Example 3.4

Consider a special case of the imperfect simply supported column of Fig. 3.9 in which the imperfection consists of a single sine half-wave of amplitude $w_1^0 = e$. It follows from (3.31) that an axial load P gives the transverse displacement

$$w(x) = \frac{e}{P_E/P - 1} \sin\left(\pi \frac{x}{\ell}\right)$$

where the Euler load was introduced in (3.21) as $P_E = (\pi/\ell)^2 EI$. The moment in the column follows from (3.24) as

$$M(x) = -EI \frac{d^2 w}{dx^2} = EI \left(\frac{\pi}{\ell}\right)^2 \frac{e}{P_E/P - 1} \sin\left(\pi \frac{x}{\ell}\right)$$

The first factors are recognized as the Euler load P_E, whereby

$$M(x) = \frac{1}{1 - P/P_E} P e \sin\left(\pi\frac{x}{\ell}\right) = \frac{M_0(x)}{1 - P/P_E}$$

This result shows, that the actual moment of a column with an initial imperfection of the shape of the first buckling mode, is obtained as the moment $M_0(x)$ of the axial load and the initial imperfection, multiplied by an amplification factor $1/(1 - P/P_E)$. This amplification factor is identical to that encountered in Example 3.1 and illustrated in Fig. 3.3.

Engineering design of columns is based on the idea that the column has an initial imperfection, and the moment generated by the amplified imperfection is then included among the section forces included in the design. A representative value of the imperfection is obtaind from experiments and usually specified in the design code. A typical value for steel columns is $e \simeq 0.003\ell$.

3.4 Bending of beam-columns

It is of interest to study the influence of an axial force in a beam on the bending stiffness. The two basic cases of symmetric and anti-symmetric bending are shown in Fig 3.10. The axial force is here considered as compressive, i.e. $P = -N \geq 0$.

Figure 3.10: Symmetric and anti-symmetric beam-column bending.

Let the beam be homogeneous of length ℓ. The differential equation (3.8) is then written in the normalized form

$$\frac{\mathrm{d}^4 w}{\mathrm{d}x^4} + k^2 \frac{\mathrm{d}^2 w}{\mathrm{d}x^2} = 0 \tag{3.32}$$

where the parameter $k^2 = P/EI$ was introduced in (3.10) and the general solution was given in (3.11).

In solving the symmetric and anti-symmetric bending problems of Fig. 3.10 it is convenient to let $x = 0$ correspond to the center of the beam. The symmetric problem will then have an even displacement function $w_s(x)$, while the anti-symmetric problem has an odd displacement function $w_a(x)$. Each of these will only contain two arbitrary constants, thereby simplifying the solution.

Symmetric bending

The symmetric bending problem has a solution of the form

$$w_s(x) = C_1 + C_3 \cos(kx) \tag{3.33}$$

The arbitrary constants C_1 and C_3 are determined from normalized boundary conditions at $x = \frac{1}{2}\ell$,

$$
\begin{aligned}
w_s(\tfrac{1}{2}\ell) &= C_1 + \cos(\tfrac{1}{2}k\ell)\, C_3 &= 0 \\
w_s'(\tfrac{1}{2}\ell) &= -k \sin(\tfrac{1}{2}k\ell)\, C_3 &= -1
\end{aligned}
\tag{3.34}
$$

These equations are solved to give

$$C_3 = \frac{1}{k \sin(\tfrac{1}{2}k\ell)} \quad , \quad C_1 = -\frac{1}{k}\cot(\tfrac{1}{2}k\ell) \tag{3.35}$$

Substitution into (3.33) then gives the displacement in the symmetric bending problem as

$$w_s(x) = \frac{\cos(kx) - \cos(\tfrac{1}{2}k\ell)}{k \sin(\tfrac{1}{2}k\ell)} \tag{3.36}$$

The moment distribution now follows from (3.7) as

$$M_s(x) = -EI\, w_s''(x) = EI\, k\, \frac{\cos(kx)}{\sin(\tfrac{1}{2}k\ell)} \tag{3.37}$$

The moment at $x = \frac{1}{2}\ell$ is of particular interest, as it gives the symmetric bending stiffness of the beam-column. It follows from (3.37) in the form

$$M_s = \frac{2EI}{\ell}(\tfrac{1}{2}k\ell)\cot(\tfrac{1}{2}k\ell) = \varphi\, \frac{2EI}{\ell} \tag{3.38}$$

where the symmetric bending stiffness coefficient φ is defined as

$$\varphi(k\ell) = (\tfrac{1}{2}k\ell)\cot(\tfrac{1}{2}k\ell) \quad , \quad \varphi_{P=0} = 1 \tag{3.39}$$

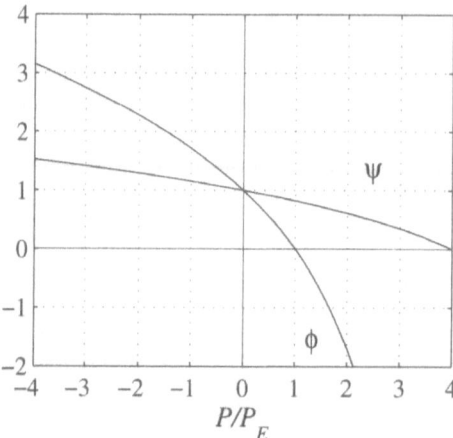

Figure 3.11: Beam-column bending coefficients φ and ψ.

The bending stiffness coefficient φ represents the change in symmetric bending stiffness of the beam due to the axial force P.

Figure 3.11 shows the bending stiffness coefficient φ as a function of the normalized axial force P/P_E, with the *Euler load* defined as

$$P_E = EI \left(\frac{\pi}{\ell}\right)^2 \qquad (3.40)$$

With $k^2 = P/EI$ it follows from (3.10) that the normalized axial load is

$$\frac{P}{P_E} = \left(\frac{k\ell}{\pi}\right)^2 \qquad (3.41)$$

It is seen that the bending stiffness vanishes for $P = P_E$, corresponding to $k\ell = \pi$. At this value of the axial load deformations can occur without a bending moment.

Anti-symmetric bending

The anti-symmetric bending displacement distribution is of the form

$$w_a(x) = C_2\, k\ell + C_4\, \sin(kx) \qquad (3.42)$$

The arbitrary constants C_2 and C_4 are determined from the boundary conditions at $x = \frac{1}{2}\ell$,

$$
\begin{aligned}
w_a(\tfrac{1}{2}\ell) &= \tfrac{1}{2}k\ell\, C_2 + \sin(\tfrac{1}{2}k\ell)\, C_4 = 0 \\
w_a'(\tfrac{1}{2}\ell) &= k\, C_2 + k\, \cos(\tfrac{1}{2}k\ell)\, C_4 = -1
\end{aligned}
\tag{3.43}
$$

Multiplication of the last equation by $\frac{1}{2}\ell$, followed by subtraction, gives

$$
\left[\, \sin(\tfrac{1}{2}k\ell) - (\tfrac{1}{2}k\ell)\cos(\tfrac{1}{2}k\ell) \,\right] C_4 = \tfrac{1}{2}\ell
\tag{3.44}
$$

Thus, the two arbitrary constants are

$$
C_4 = \frac{\tfrac{1}{2}\ell}{\sin(\tfrac{1}{2}k\ell) - (\tfrac{1}{2}k\ell)\cos(\tfrac{1}{2}k\ell)} \quad , \quad
C_2 = -\frac{\sin(\tfrac{1}{2}k\ell)}{\tfrac{1}{2}k\ell}\, C_4
\tag{3.45}
$$

The displacement distribution in the anti-symmetric bending problem is then obtained as

$$
w_a(x) = \frac{\tfrac{1}{2}\ell\, \sin(kx) - x\, \sin(\tfrac{1}{2}k\ell)}{\sin(\tfrac{1}{2}k\ell) - (\tfrac{1}{2}k\ell)\cos(\tfrac{1}{2}k\ell)}
\tag{3.46}
$$

The moment distribution follows by differentiation as

$$
M_a(x) = -EI\, w_a''(x) = EI\, k\, \frac{\tfrac{1}{2}k\ell\, \sin(kx)}{\sin(\tfrac{1}{2}k\ell) - (\tfrac{1}{2}k\ell)\cos(\tfrac{1}{2}k\ell)}
\tag{3.47}
$$

The moment at $x = \frac{1}{2}\ell$ gives the anti-symmetric bending stiffness. In the present case it takes the form

$$
M_a = \frac{6EI}{\ell}\, \frac{\tfrac{1}{12}(k\ell)^2}{1 - (\tfrac{1}{2}k\ell)\cot(\tfrac{1}{2}k\ell)} = \psi\, \frac{6EI}{\ell}
\tag{3.48}
$$

where the anti-symmetric bending stiffness coefficient ψ is defined as

$$
\psi(k\ell) = \frac{\tfrac{1}{12}(k\ell)^2}{1 - (\tfrac{1}{2}k\ell)\cot(\tfrac{1}{2}k\ell)} \quad , \quad \psi_{P=0} = 1
\tag{3.49}
$$

The anti-symmetric bending stiffness coefficient ψ represents the change of bending stiffness due to the axial force. It is conveniently expressed as a function of the normalized axial force P/P_E as shown in Fig. 3.11. It is seen that in the anti-symmetric bending problem the bending stiffness vanishes at $P = 4P_E$, corresponding to the second buckling mode.

Beam-column bending element

The stiffness properties of a beam are conveniently expressed by the element stiffness matrix as explained in detail in Section 2.6 for simple beam bending. The stiffness matrix $\mathbf{K}$ gives the nodal force vector $\mathbf{f}^T = [\, P_1, M_1, P_2, M_2\,]$ in terms of the nodal displacement vector $\mathbf{u}^T = [\, w_1, \theta_1, w_2, \theta_2\,]$. The standard method of deriving the stiffness matrix is to start out from a representation of the displacement field in terms of the nodal displacement components, i.e. a representation of the form

$$w(x) = N_1(x)w_1 + N_2(x)\theta_1 + N_3(x)w_2 + N_4(x)\theta_2 \qquad (3.50)$$

where the shape functions $N_1(x), \cdots, N_4(x)$ are solutions to the governing differential equation, in the present case (3.9). The internal force $Q(x)$ and moment $M(x)$ can then be obtained by differentiation as given in (3.12) and (3.13), and the nodal forces obtained from the end point values $Q(\pm\frac{1}{2}\ell)$ and $M(\pm\frac{1}{2}\ell)$. The following derivation corresponds to compression, i.e. $P \geq 0$.

The first task is to find the shape functions. The shape functions $N_2(x)$ and $N_4(x)$ correspond to rotations, without translation of the end points. They can be expressed directly by the sum and difference of the symmetric and anti-symmetric deformation functions $w_s(x)$ and $w_a(x)$, determined above.

$$N_2(x) = \tfrac{1}{2}[\, w_a(x) - w_s(x)\,] \quad , \quad N_4(x) = \tfrac{1}{2}[\, w_a(x) + w_s(x)\,] \quad (3.51)$$

It is seen directly from Fig. 3.10 that $N_2(x)$ corresponds to the unit rotation $\theta_1 = 1$, and $N_4(x)$ corresponds to the unit rotation $\theta_2 = 1$.

Figure 3.12: Beam-column deformation modes for w_1 and w_2.

The shape functions $N_1(x)$ and $N_3(x)$, corresponding to unit translations $w_1 = 1$ and $w_2 = 1$, respectively, are shown in Fig 3.12. They consist of a linear part plus or minus a contribution from anti-symmetric bending. In the present formulation $x = 0$ is at the center of the beam, and these shape functions are

$$\begin{aligned}
N_1(x) &= \frac{1}{2}\Big(1 - \frac{2x}{\ell}\Big) - \frac{1}{\ell}w_a(x) \\
N_3(x) &= \frac{1}{2}\Big(1 + \frac{2x}{\ell}\Big) + \frac{1}{\ell}w_a(x)
\end{aligned} \qquad (3.52)$$

Box 3.3: Stiffness matrix for beam-column bending

The stiffness matrix K of a beam-column of length ℓ, bending stiffness EI, and axial compression force P is expressed in terms of the symmetric and anti-symmetric bending stiffness coefficients φ and ψ.

In compression $(P \geq 0)$ the parameter k is defined by $k^2 = P/EI$, and the bending coefficients are

$$\varphi = (\tfrac{1}{2}k\ell)\cot(\tfrac{1}{2}k\ell) \quad , \quad \psi = \frac{\tfrac{1}{12}(k\ell)^2}{1 - (\tfrac{1}{2}k\ell)\cot(\tfrac{1}{2}k\ell)}$$

In tension $(P \leq 0)$ the parameter k is defined by $k^2 = -P/EI$, and the bending coefficients are

$$\varphi = (\tfrac{1}{2}k\ell)\coth(\tfrac{1}{2}k\ell) \quad , \quad \psi = -\frac{\tfrac{1}{12}(k\ell)^2}{1 - (\tfrac{1}{2}k\ell)\coth(\tfrac{1}{2}k\ell)}$$

The coefficients are normalized such that $\varphi=\psi=1$ for $P=0$.

The general expression for the stiffness matrix is

$$\mathsf{K} = \frac{EI}{\ell^3} \begin{bmatrix} 12\,\psi\,\varphi & -6\,\psi\,\ell & -12\,\psi\,\varphi & -6\,\psi\,\ell \\ -6\,\psi\,\ell & (3\psi + \varphi)\ell^2 & 6\,\psi\,\ell & (3\psi - \varphi)\ell^2 \\ -12\,\psi\,\varphi & 6\,\psi\,\ell & 12\,\psi\,\varphi & 6\,\psi\,\ell \\ -6\,\psi\,\ell & (3\psi - \varphi)\ell^2 & 6\,\psi\,\ell & (3\psi + \varphi)\ell^2 \end{bmatrix}$$

Thus, the functional form has already been calculated in (3.15) and (3.25).

The shear force $Q_j(x)$ and internal moment $M_j(x)$ corresponding to the shape function $N_j(x)$ follows from (3.4) and (3.7) as

$$Q_j(x) = -EI[\,N_j'''(x) + k^2\,N_j'(x)\,] \quad , \quad M_j(x) = -EI\,N_j''(x) \quad (3.53)$$

These formulae are now used to calculate $\pm Q_j(\pm\tfrac{1}{2}\ell)$ and $\pm M_j(\pm\tfrac{1}{2}\ell)$, and these values form the stiffness matrix given in Box 3.3. A complete stiffness matrix, including the extension of the beam-column as well as initial imperfections is derived in Krenk et al. (1999).

Alternatively, the stiffness matrix can be obtained directly from the symmetric and anti-symmetric bending moment relations (3.17) and (3.27) in combination with the equilibrium equations. The bending cases $\theta_1 = 1$ and $\theta_2 = 1$ follow directly from superposition, as indicated in the shape function formulae (3.30). In the translation cases $w_1 = 1$ and $w_2 = 1$, shown in Fig. 3.12, it is important to include the moment contributions $w_1 P$ and $-w_2 P$, respectively, when evaluating the transverse forces.

The stiffness matrix for axial tension, $P \leq 0$, can be obtained directly from the expressions obtained for compression by introducing the following changes. The parameter k is now defined by

$$k^2 = -\frac{P}{EI} \geq 0 \tag{3.54}$$

and in the expressions k is replaced by the imaginary number ik. This leads to the bending stiffness coefficients defined in Box 3.3, where trigonometric functions are replaced by their hyperbolic equivalents, and k^2 is replaced by $(ik)^2 = -k^2$. In some high level computer languages like MATLAB these replacements may be performed automatically via complex calculus by using the definition $k = \sqrt{P/EI}$, also for $P < 0$.

Example 3.5

Figure 3.13 shows a simply supported column of length ℓ loaded by an axial compression force P and a bending moment M acting at the right end. The purpose is to determine the bending stiffness of the column with respect to this moment, first by traditional analysis, and then by a simple symmetry argument.

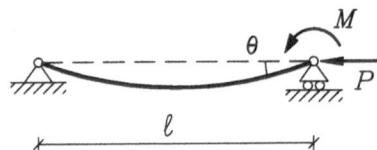

Figure 3.13: Bending of simply supported beam-column.

The general solution to the homogeneous column problem is given in (3.11),

$$w(x) = C_1 + C_2\,kx + C_3\,\cos(kx) + C_4\,\sin(kx)$$

Let the origin be at the unloaded left support. The conditions $w(0) = 0$ and $w''(0)$ then leads to $C_1 = C_3 = 0$ as shown in Section 3.2. The boundary conditions at the right end then are

$$w(\ell) = k\ell\, C_2 + \sin(k\ell)\, C_4 = 0$$
$$w''(\ell) = -k^2 \sin(k\ell)\, C_4 = -M/EI$$

Solution of these equations gives

$$C_4 = \frac{1}{k^2 \sin(k\ell)} \frac{M}{EI} \quad , \quad C_2 = \frac{-1}{k^2\,(k\ell)} \frac{M}{EI}$$

The rotation of the right end is

$$\theta = -w'(\ell) = -C_2\, k - C_4\, k\, \cos(k\ell) = \left(\frac{1}{(k\ell)^2} - \frac{\cot(k\ell)}{k\ell}\right) \frac{\ell M}{EI}$$

From this the stiffness relation follows in the form

$$M = \frac{(k\ell)^2}{1 - k\ell\, \cot(k\ell)} \frac{EI}{\ell}\, \theta$$

This is a standard derivation of the stiffness by solving the differential equation.

In the present case the stiffness can be obtained directly by observing that the boundary conditions of the column in Fig. 3.13 correspond to the left half of the anti-symmetric bending case shown in Fig. 3.10. The moment-rotation relation of the present case therefore follows directly from (3.48) by replacing the length $\frac{1}{2}\ell$ of the anti-symmetric bending problem with ℓ in the current problem. Thus, the rescaled version of (3.48) is

$$M = \psi(2k\ell)\frac{3EI}{\ell}$$

This reproduces the result obtained above by integration of the differential equations. Note, that the limit $P = 0$ reproduces the beam bending result $M = 3EI/\ell$ from (A.4) in Appendix A.

Example 3.6

Figure 3.14a shows a column of length 2ℓ with simple supports at the ends and a transverse spring support with spring constant c at the center. The column can always buckle in the anti-symmetric mode shown in Fig. 3.14c with critical load is $P_E = EI(\pi/\ell)^2$, independent of the spring stiffness. There is also the possibility of symmetric modes of buckling

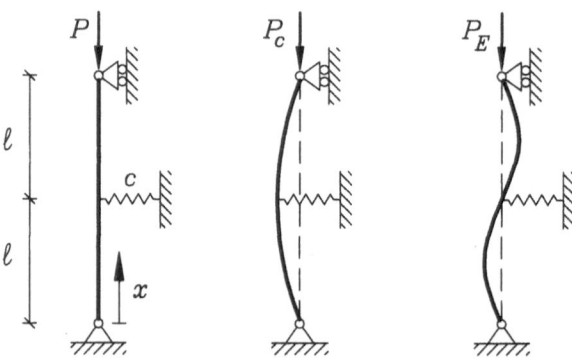

Figure 3.14: Buckling of column with transverse spring support.

as shown in Fig. 3.14b. The critical load in these modes depend on the spring stiffness. In fact, for very large spring stiffness the center appears as fixed, and the symmetric buckling problem corresponds to that of Example 3.3 with $P_c \simeq 2.05 P_E$, while for very small spring stiffness the center appears as unsupported leading to an Euler column of length 2ℓ and critical load $P_c = \frac{1}{4} P_E$. Thus, for some finite value of the spring stiffness the lowest critical mode of buckling changes from symmetric to anti-symmetric, and further increase of the spring stiffness does not increase the overall stability of the column. This example sets up the general stability equation of the symmetric modes and determines the value c_0 of the spring stiffness, where the symmetric and anti-symmetric buckling mode both correspond to the same critical load P_E.

Introduce a coordinate system with x-from the bottom, as shown in Fig. 3.14a. The symmetric modes can then be analyzed by considering only the lower half as illustrated in Fig. 3.15. At the left end $w(0) = 0$ and $w''(0) = 0$. As demonstrated in the previous example the solution

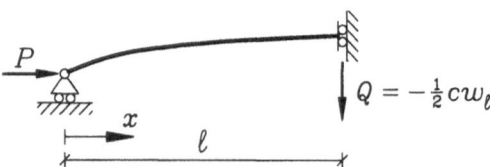

Figure 3.15: Lower half of the spring supported column from Fig. 3.14.

then takes the form

$$w(x) = C_2\,kx + C_4\sin(kx)$$

The arbitrary constants C_2 and C_4 are to be determined by the boundary conditions at $x = \ell$.

It follows from symmetry, that there is no rotation at $x = \ell$, and thus

$$w'(\ell) = k\,C_2 + k\cos(k\ell)\,C_4 = 0$$

The other boundary condition at $x = \ell$ prescribes that the shear force corresponds to *half* the spring force,

$$Q(\ell) = -\tfrac{1}{2}\,c\,w_\ell$$

Substitution of the shear force from (3.13), $Q = -EIk^3C_2$, and the displacement w_ℓ leads to the following homogeneous form of the static boundary condition

$$\frac{1}{EI}[Q + \tfrac{1}{2}c\,w_\ell] = [-k^3 + \frac{c}{2EI}\,k\ell]\,C_2 + \frac{c}{2EI}\sin(k\ell)\,C_4 = 0$$

The two homogeneous boundary conditions can be arranged into the following matrix format

$$\begin{bmatrix} 1 & \cos(k\ell) \\ \dfrac{c\,\ell^3}{2EI} - (k\ell)^2 & \dfrac{c\,\ell^3}{2EI}\dfrac{\sin(k\ell)}{k\ell} \end{bmatrix} \begin{bmatrix} C_2 \\ C_4 \end{bmatrix} = \begin{bmatrix} 0 \\ 0 \end{bmatrix}$$

In order to permit a non-trivial solution the determinant of this equation system must vanish, giving the equation

$$\frac{c\,\ell^3}{2EI}\frac{\sin(k\ell)}{k\ell} - \left[\frac{c\,\ell^3}{2EI} - (k\ell)^2\right]\cos(k\ell) = 0$$

For $c = 0$ this equation reduces to $\cos(k\ell) = 0$ with the roots $k\ell = \frac{1}{2}\pi, \frac{3}{2}\pi, \cdots$, corresponding to the even buckling modes of an Euler column of length 2ℓ.

For positive spring stiffness the equation is written in the form

$$\tan(k\ell) = k\ell - \frac{2EI}{c\,\ell^3}\,(k\ell)^3$$

This equation is illustrated in Fig. 3.16, showing the left side and the right side for different values of the spring stiffness c. For infinite spring

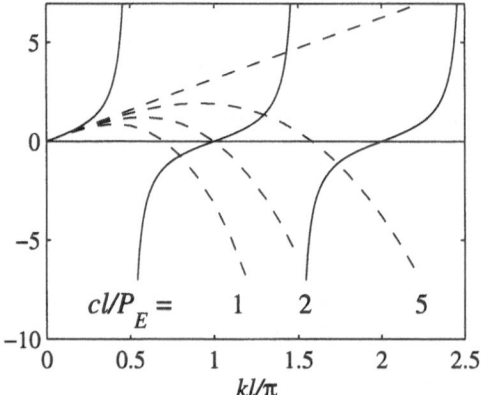

Figure 3.16: Stability equation: $\tan(k\ell)$ (–) and $k\ell - 2\dfrac{EI}{c\ell^3}(k\ell)^3$ (- -).

stiffness the right side is simply $k\ell$, and the figure corresponds to Fig. 3.8 from Example 3.3. For finite spring stiffness a cubic term is subtracted, and the curve dips, leading to intersection with the tangent curves at lower values of $(k\ell)_n$.

The general solution requires iterations, but the present question of co-inciding critical load of value P_E for the lowest symmetric and anti-symmetric buckling mode can be answered in closed form. Indeed, this critical load corresponds to $(k\ell)_1 = \pi$, and substitution of this value into the transcendental equation leads to

$$0 = \pi - \frac{2EI}{c_0\,\ell^3}\pi^3$$

This determines the corresponding spring stiffness

$$c_0 = 2\pi^2\frac{EI}{\ell^3} = 2\frac{P_E}{\ell}$$

It is also seen form Fig. 3.16, that the curve corresponding to this value intersects the tangent curve on the axis at $k\ell/\pi = 1$.

3.5 The 'Elastica'

The column and beam-column problems studied so far in this chapter have been based on the linearized theory presented in Section 3.1 and summarized in Box 3.1. This theory fits well within the usual methods of structural analysis and can account for most problems of elastic columns with sufficient accuracy for practical applications. However, it is interesting to study the general, fully non-linear solution of the column under a force applied at the end. It was this fully nonlinear problem of 'the elastica' that was studied by LEONHARD EULER (1644), and Fig. 3.17 is one of the illustrations from his treatise, clearly showing large deformations. The solution presented in this section was obtained by KIRCHHOFF (1859) making use of elliptic integrals, see e.g. Bažant & Cedolin (1991).

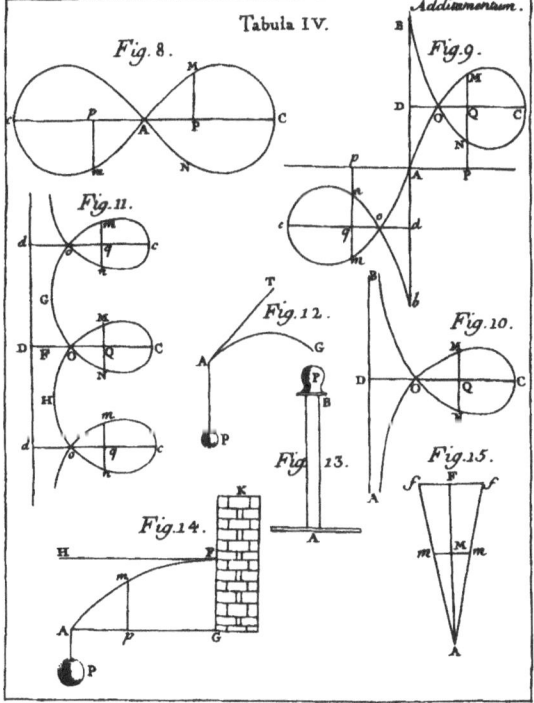

Figure 3.17: LEONHARD EULER *De curvis elasticis* (1644).

Differential equation

The problem of the elastica is here studied with reference to the column problem shown in Fig 3.18. A vertical force P is applied to an inextensible elastic column, in which the bending moment M is proportional to the local curvature κ. The curvature is defined as the derivative of the angle of the tangent θ wih respect to the length s along the column. Thus, the kinematic and constitutive relations of the theory are

$$\frac{d\theta}{ds} = \kappa \quad , \quad \frac{M}{EI} = \kappa \qquad (3.55)$$

where the bending stiffness EI is assumed constant in the present application.

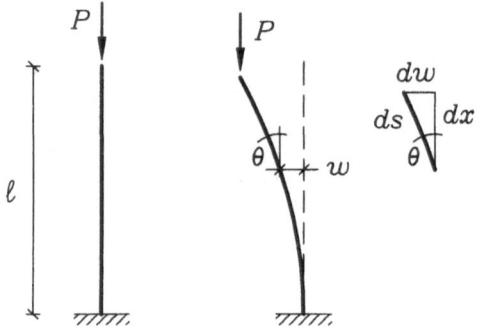

Figure 3.18: Finite deflection of end-loaded column.

In the problem shown in Fig. 3.18 the moment M at any section follows from statics as

$$M(s) = P[w_\ell - w(s)] \qquad (3.56)$$

Differentiation with respect to the arc length s gives

$$\frac{dM}{ds} = -P\frac{dw}{ds} \qquad (3.57)$$

The derivative on the right is expressed in terms of the angle θ by use of (3.55), and the derivative on the left follows from the detail in Fig. 3.18 as $dw/ds = \sin\theta$. Thus, the differential equation for finite deflection of a homogeneous elastic column is obtained in the form

$$EI\frac{d^2\theta}{ds^2} = -P\sin\theta \qquad (3.58)$$

As in the linear case it is convenient to introduce the parameter

$$k^2 = \frac{P}{EI} \tag{3.59}$$

When introducing the buckling load

$$P_0 = \left(\frac{\pi}{2\ell}\right)^2 EI \tag{3.60}$$

the parameter k^2 can be expressed as

$$k^2 = \left(\frac{\pi}{2\ell}\right)^2 \frac{P}{P_0} \tag{3.61}$$

In terms of the parameter k^2 the finite deflection column equation becomes

$$\frac{d^2\theta}{ds^2} + k^2 \sin\theta = 0 \tag{3.62}$$

This equation is similar to that governing oscillations of a pendulum under gravity, and in both cases the linearized equation is obtained by replacing $\sin\theta$ with θ.

The differential equation (3.62) can be integrated by a method inspired by the pendulum analogy. An energy integral of a conservative equation of motion can be obtained by multiplication with velocity. In the present case this corresponds to multiplication of (3.62) by $d\theta/ds$. The equation can then be integrated by use of the differentiation relations

$$\frac{1}{2}\frac{d}{ds}\left(\frac{d\theta}{ds}\right)^2 = \frac{d^2\theta}{ds^2}\frac{d\theta}{ds} \tag{3.63}$$

and

$$2\frac{d}{ds}\left(\sin\left(\tfrac{1}{2}\theta\right)\right)^2 = 2\sin\left(\tfrac{1}{2}\theta\right)\cos\left(\tfrac{1}{2}\theta\right)\frac{d\theta}{ds} = \sin\theta\,\frac{d\theta}{ds} \tag{3.64}$$

When using these relations from right to left, the following integrated form of (3.62) is obtained,

$$\frac{1}{4k^2}\left(\frac{d\theta}{ds}\right)^2 = c^2 - \sin^2\left(\tfrac{1}{2}\theta\right) \tag{3.65}$$

where c^2 is an arbitrary constant, to be determined by the boundary conditions.

At the loaded end of the elastica, $s = \ell$, the moment vanishes, and therefore also the curvature vanishes, $d\theta/ds = 0$. When this condition is substituted into (3.65), the arbitrary constant c is found to be

$$c = \sin\left(\tfrac{1}{2}\theta_\ell\right) \tag{3.66}$$

Thus, the parameter c measures the magnitude of the deformation via the rotation of the point of application of the force.

Integration in terms of elliptic integrals

Taking the square root of the differential equation (3.65) with a positive sign, and thereby considering displacements to the left as shown in Fig. 3.18, the differential equation takes the form

$$\frac{d\theta}{\sqrt{c^2 - \sin^2\left(\tfrac{1}{2}\theta\right)}} = 2k \, ds \tag{3.67}$$

This relation leads to elliptic integrals, when the following substitution is made,

$$\sin\left(\tfrac{1}{2}\theta\right) = c \sin\varphi \tag{3.68}$$

Differentiation of this formula gives

$$\tfrac{1}{2}\cos\left(\tfrac{1}{2}\theta\right) d\theta = c \cos\varphi \, d\varphi \tag{3.69}$$

and substitution into (3.67) then leads to

$$k\,ds = \frac{\tfrac{1}{2}\,d\theta}{c\cos\varphi} = \frac{d\varphi}{\cos\left(\tfrac{1}{2}\theta\right)} = \frac{d\varphi}{\sqrt{1 - c^2\sin^2\varphi}} \tag{3.70}$$

It follows from the substitution (3.68) and the boundary conditions that the φ varies between 0 at $s = 0$ and $\tfrac{1}{2}\pi$ at $s = \ell$. Thus, the parameter $k\ell$ can be obtained by integration of (3.70).

$$k\ell = \int_0^{\pi/2} \frac{d\varphi}{\sqrt{1 - c^2\sin^2\varphi}} \tag{3.71}$$

The integral on the right is called the complete elliptic integral of the first kind, and is denoted

$$K(c) = \int_0^{\pi/2} \frac{d\varphi}{\sqrt{1 - c^2\sin^2\varphi}} \tag{3.72}$$

according to the notation of e.g. Gradshteyn & Ryzhik (1980). Slightly different notations are in use, and the notation should be carefully checked before using existing software. Elliptic integrals and functions are available in high level programs like MAPLE and MATLAB. When the parameter c is introduced from (3.66) and the load ratio P/P_0 from (3.61), the load is obtained

$$\sqrt{\frac{P}{P_0}} = \frac{2}{\pi} K[\sin(\tfrac{1}{2}\theta_\ell)] \tag{3.73}$$

giving the load in terms of the rotation θ_ℓ.

The transverse displacement w_ℓ of the top of the elastica can be determined from (3.65) by using that $\theta_0 = 0$, whereby the parameter c becomes

$$c = \frac{1}{2k}\frac{d\theta_0}{ds} = \frac{1}{2k}\frac{M_0}{EI} = \frac{w_\ell}{2k}\frac{P}{EI} = \tfrac{1}{2} k w_\ell \tag{3.74}$$

The parameters c and $k\ell$ were expressed in terms of the rotation θ_ℓ in (3.66) and (3.71), respectively, and upon substitution of these expressions

$$\frac{w_\ell}{\ell} = \frac{2\sin\left(\tfrac{1}{2}\theta_\ell\right)}{K[\sin(\tfrac{1}{2}\theta_\ell)]} \tag{3.75}$$

giving the transverse displacement w_ℓ as function of the rotation θ_ℓ.

The vertical displacement of the top of the elastica is found by integration of the relation

$$\frac{dx}{ds} = \cos\theta \tag{3.76}$$

illustrated in the detail in Fig. 3.18. This differential relation can be integrated in the following way. First $\cos\theta$ is reduced to half angle by a standard trigonometric relation,

$$dx = \cos\theta\, ds = [2\cos^2(\tfrac{1}{2}\theta) - 1]\, ds \tag{3.77}$$

and then ds is substituted from (3.70), giving

$$dx + ds = 2\cos^2(\tfrac{1}{2}\theta)\, ds = \frac{2}{k}\cos(\tfrac{1}{2}\theta)\, d\varphi \tag{3.78}$$

Integration of this relation over the full length of the elastica gives

$$(\ell - u_\ell) + \ell = \frac{2}{k}\int_0^{\pi/2} \sqrt{1 - c^2\sin^2\varphi}\, d\varphi \tag{3.79}$$

where u_ℓ denotes the downward displacement of the end of the elastica. The integral is called the complete elliptic integral of the second kind, and is denoted, Gradshteyn & Ryzhik (1980),

$$E(c) = \int_0^{\pi/2} \sqrt{1 - c^2 \sin^2 \varphi} \, d\varphi \qquad (3.80)$$

When using the expression (3.71) for $k\ell$, the final form of the axial displacement of the top of the elastica becomes

$$\frac{u_\ell}{\ell} = 2\left(1 - \frac{E[\sin(\frac{1}{2}\theta_\ell)]}{K[\sin(\frac{1}{2}\theta_\ell)]}\right) \qquad (3.81)$$

This completes the description of the magnitude of the load and the displacement of the point of load application in terms of the parameter θ_ℓ. Simple asymptotic formulas, valid for moderate displacements, are obtained after deriving expressions for the deformed shape of the elastica in the following section.

The deformed shape of the elastica

In order to determine the deformed shape of the elastica it is convenient first to establish a relation between the lenght coordinate s along the elastica and the angle φ. This relation follows from integration of (3.70),

$$k s = \int_0^\varphi \frac{d\tilde{\varphi}}{\sqrt{1 - c^2 \sin^2 \tilde{\varphi}}} \qquad (3.82)$$

The integral is similar to (3.71) but with an upper limit in the interval from 0 to $\frac{1}{2}\pi$. This integral is called the incomplete elliptic integral of the first kind, Gradshteyn & Ryzhik (1980),

$$F(\varphi, c) = \int_0^\varphi \frac{d\tilde{\varphi}}{\sqrt{1 - c^2 \sin^2 \tilde{\varphi}}} \qquad (3.83)$$

Note in particular that $F(\frac{1}{2}\pi, c) = K(c)$. When use is made of the expression (3.71) for $k\ell$,

$$\frac{s}{\ell} = \frac{F[\varphi, \sin(\frac{1}{2}\theta_\ell)]}{K[\sin(\frac{1}{2}\theta_\ell)]} \qquad (3.84)$$

This is a relation between the arc-length s and the angle φ for any particular value of the rotation θ_ℓ. In view of this relation the angle φ may be used as parameter in lieu of the arc-length s, when describing the deformed shape of the elastica.

Let the deformed shape of the elastica be represented in parametric form as $[x(s), y(s)]$, or the equivalent form $[x(\varphi), y(\varphi)]$. The coordinate functions $x(s)$ and $y(s)$ then satisfy the differential relations

$$\frac{dy}{ds} = \sin\theta \quad , \quad \frac{dx}{ds} = \cos\theta \tag{3.85}$$

as illustrated in the detail in Fig. 3.18. These differential equations are integrated by substituting the variable φ, using that $k\,ds = d\varphi / \cos(\frac{1}{2}\theta)$. By this substitution the first differential equation takes the form

$$dy = \frac{1}{k}\frac{\sin\theta}{\cos(\frac{1}{2}\theta)}\,d\varphi = \frac{2}{k}\sin(\tfrac{1}{2}\theta)\,d\varphi = \frac{2c}{k}\sin\varphi\,d\varphi \tag{3.86}$$

This equation integrates directly to

$$y = \frac{2c}{k}\left[1 - \cos\varphi\right] \tag{3.87}$$

and by introduction of the expressions for $k\ell$ and for c,

$$\frac{y}{\ell} = \frac{2\sin(\frac{1}{2}\theta_\ell)}{K[\sin(\frac{1}{2}\theta_\ell)]}\left[1 - \cos\varphi\right] \tag{3.88}$$

where the first factor is recognized as the transverse displacement w_ℓ at the top of the elastica.

The second of the differential equations (3.85) was already reduced in (3.76)-(3.78) to the form

$$dx + ds = \frac{2}{k}\cos(\tfrac{1}{2}\theta)\,d\varphi = \frac{2}{k}\sqrt{1 - c^2\sin^2\varphi}\,d\varphi \tag{3.89}$$

When the incomplete elliptic integral of the second kind is introduced as, Gradshteyn & Ryzhik (1980),

$$E(\varphi, c) = \int_0^\varphi \sqrt{1 - c^2\sin^2\tilde\varphi}\,d\tilde\varphi \tag{3.90}$$

this differential equation is integrated to give

$$\frac{x}{\ell} + \frac{s}{\ell} = 2\frac{E[\varphi, \sin(\frac{1}{2}\theta_\ell)]}{K[\sin(\frac{1}{2}\theta_\ell)]} \tag{3.91}$$

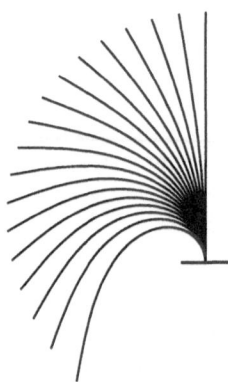

Figure 3.19: Deformed elastica for $\theta_\ell = 0°, 10°, \cdots, 170°$.

Finally, substitution of s/ℓ from (3.84) then leads to

$$\frac{x}{\ell} = \frac{2\,E[\,\varphi, \sin(\tfrac{1}{2}\theta_\ell)]\; -\; F[\,\varphi, \sin(\tfrac{1}{2}\theta_\ell)]}{K[\,\sin(\tfrac{1}{2}\theta_\ell)]} \qquad (3.92)$$

The formula (3.81) for the vertical displacement at the end of the elastica is a special case of this general result.

Figure 3.19 illustrates the increasingly deformed shape of the elastica. Note the similarity with the special cases of EULER's illustrations in Fig. 3.17.

Asymptotic load-displacement relations

While the above derivation gives the solution for any magnitude of load and displacement, it is usually sufficient to know the solution to within the first correction of the linear solution. This asymptotic solution - for small but finite displacements - is derived below.

For small displacements the parameter c is small and the following asymptotic series expansions can be obtained by replacing the integrand of the elliptic integrals (3.72) and (3.80) by their Taylor expansion.

$$K(c) \simeq \frac{\pi}{2}\Big(1 + \tfrac{1}{4}c^2 + \cdots\Big) \quad , \quad E(c) \simeq \frac{\pi}{2}\Big(1 - \tfrac{1}{4}c^2 - \cdots\Big) \qquad (3.93)$$

In the following only the first term is needed in the representation of c, i.e. $c \simeq \tfrac{1}{2}\theta_\ell$.

Asymptotic expressions are now obtained for the load P, the transverse displacement w_ℓ, and the axial displacement u_ℓ in terms of the rotation θ_ℓ. Subsequently the rotation is eliminated to obtain direct asymptotic relations between P, w_ℓ and u_ℓ.

The asymptotic relation (3.93a) is substituted into (3.73) to give

$$\sqrt{P/P_0} \simeq 1 + \tfrac{1}{4}c^2 + \cdots \simeq 1 + \tfrac{1}{16}\theta_\ell^2 + \cdots \qquad (3.94)$$

and by taking the square of the series and retaining only the two first terms

$$P/P_0 \simeq 1 + \tfrac{1}{8}\theta_\ell^2 + \cdots \qquad (3.95)$$

The asymptotic relation for w_ℓ is obtained directly by writing (3.75) in the form

$$\frac{w_\ell}{\ell} = \frac{2c}{\frac{\pi}{2}\sqrt{P/P_0}} \simeq \frac{2}{\pi}\theta_\ell\left(1 + \cdots\right) \qquad (3.96)$$

Finally, the asymptotic relation for u_ℓ follows from (3.81) by substitution of (3.93),

$$\frac{u_\ell}{\ell} \simeq 2\left[1 - (1 - \tfrac{1}{4}c^2 + \cdots)(1 - \tfrac{1}{4}c^2 + \cdots)\right] \simeq 2\frac{2}{4}c^2 \cdots \simeq \tfrac{1}{4}\theta_\ell^2 \cdots \quad (3.97)$$

It is seen that w_ℓ is approximately proportional to θ_ℓ, and it is therefore convenient to express P/P_0 and u_ℓ in terms of w_ℓ.

It follows from (3.95) by substitution of (3.96) that the load is given asymptotically by

$$\frac{P}{P_0} \simeq 1 + \frac{\pi^2}{8}\left(\frac{w_\ell}{2\ell}\right)^2 \qquad (3.98)$$

In a similar way asymptotic relations for u_ℓ are

$$\frac{u_\ell}{\ell} \simeq 2\left(\frac{P}{P_0} - 1\right) \simeq \frac{\pi^2}{4}\left(\frac{w_\ell}{2\ell}\right)^2 \qquad (3.99)$$

The asymptotic load-displacement relations (3.98) and (3.92) are shown in Fig. 3.20. It is seen that for the inextensible elastica there is no deformation until the load P reaches the stability load P_0. At this load the elastica buckles, and develops increasing deformations for increasing load. The increase of the load after buckling can be characterized by the stiffness $dP/du_\ell \simeq \tfrac{1}{2}P_0/\ell$. This stiffness is much smaller than the axial elastic stiffness of a typical column, and thus in practice the increase of the load after buckling is most

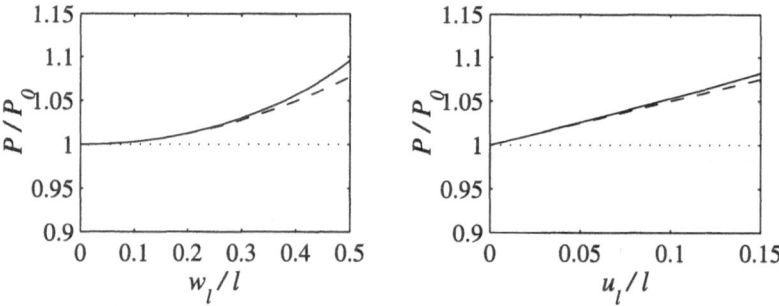

Figure 3.20: Load-displacement curves for inextensible elastica. Exact theory (–), asymptotic relations (- -), and linear theory (· · ·).

often negligible. It is however of practical interest that the column retains its load carrying capacity for loads around the buckling load, while e.g. shell structures generally exhibit buckling deformation modes that lead to a dramatic loss of load carrying capacity at buckling. In practice this means that most shell structures can only carry a fraction of their ideal buckling load due to the effect of unavoidable imperfections. A discussion of sensitivity to imperfections can be found e.g. in Thompson & Hunt (1973) and Dym (1974).

3.6   Summary

The classical theory for moderate deformation of beam-columns is obtained from the beam theory of Chapter 2 by including the effect of a normal force as an equivalent transverse loading term, proportional to the curvature of the beam-column, as shown in Box 3.1. The effect of this term is to reduce the stiffness, when the normal force is compression, while increasing the stiffness for a normal tension force. In practice the theory is mainly used to account for the reduced stiffness in the presence of compression, but the theory of shallow cables, described in Chapter 4, can be considered as a limiting case of tension in a beam-column with vanishing bending stiffness.

When a beam carries a transverse load the reduced stiffness produced by a compression force will lead to increased deflection, and the increased deflection in turn leads to larger effect of the normal force. The net result is the so-called amplification of the bending moments, illustrated by a simple case

in Example 3.1. For sufficiently large compression force there is no stiffness
left. In the case of ideally straight beam-columns this so-called *critical load*
P_c can be determined directly, without reference to any transverse load. In
this case, where the only load is the axial force, the beam-column is simply
called a column. The analysis procedure takes the form of an eigenvalue problem, with the critical load P_c as the eigenvalue, as described in Section 3.2.
Representative examples are shown in Box 3.2.

In practice, columns are not ideally straight, and the 'real-life' column problem consists in rapidly increasing deformations, when the load approaches the
critical value P_c. This problem of geometrically imperfect columns is considered in Section 3.3, where it is demonstrated, that the effect of the normal
force is to amplify the initial out of straightness. This approach forms the
basis of practical column design.

The stiffness reducing effect of the normal force is also present when beams
are joined to form frames. Usually the analysis of frames is carried out
by formulating a model in terms of beam elements, as briefly indicated in
Section 2.6. A beam-column bending element, that incorporates the effect of
the normal force, is derived in Section 3.4, and the result is summarized in
Box 3.3.

Finally, the fully non-linear column equation is solved for the case of a vertical column, clamped at its base and carrying a vertical load at its top.
The analysis is carried out in terms of elliptic integrals. It serves a triple
purpose, by illustrating the solution of a non-linear problem, providing an
accurate reference solution for non-linear finite element formulations, and by
providing simple asymptotic results for the load and displacements just after
the onset of buckling. As it turns out, the theoretical load increases slightly
beyond the critical load, but it is accompanied by so large deformation, that
the theoretical excess capacity is not usable in practice. However, the theoretical excess capacity implies that columns are not particularly sensitive to
geometric imperfections, and the critical load P_c, determined form a simple
linear eigenvalue analysis, can therefore serve as reference for evaluating the
design load, when due consideration is given to deformation and material
capacity.

3.7   Exercises

Exercise 3.1   Consider the beam-column of Example 3.1, shown in Fig. 3.3. Show that for $N = 0$ the moment at the center is $M_c^0 = (\ell/\pi)^2 p_0$. Find the moment amplification factor M_c/M_c^0 at the center, when M_c corresponds to the normal force $N \neq 0$.

Exercise 3.2   Consider a homogeneous, simply supported beam-column of length ℓ with constant normal force N and uniformly distributed transverse load of intensity p_0. Find the transverse displacement $w(x)$ by solving the differential equation (3.8),

$$EI\, w'''' - N\, w'' = p_0$$

with boundary conditions $w = 0$ and w'' at both ends. (Verify the solution for $N = 0$ by comparison with A.1.2 in Appendix A). Determine the amplification factor w_c/w_c^0 for the transverse displacement at the center, and identify the axial load $N = -P_c$ for which the displacement becomes infinite.

Exercise 3.3   Combine the theoretical developments of Sections 2.9 and 3.1 to show that the vibrations of a homogeneous beam-column with normal force N and angular frequency ω are governed by the differential equation

$$EI\, \tilde{w}'''' - N\, \tilde{w}'' - \omega^2 \rho A\, \tilde{w} = 0$$

where EI is the bending stiffness, A the cross-section area, and ρ the mass density.

Exercise 3.4   Use the differential equation from Exercise 3.3 to determine the lowest vibration frequency ω_1 of a simply supported homogeneous column of length ℓ, with bending stiffness EI, cross-section area A, mass density ρ, and normal compression force $P = -N$. Show that the result can be expressed in the form

$$\frac{\omega_1}{\omega_0} = \sqrt{1 - \frac{P}{P_E}}$$

where $P_E = (\pi/\ell)^2 EI$, and $\omega_0 = (\pi/\ell)^2 (EI/\rho A)^{1/2}$ is the natural angular frequency for $P = 0$, determined in Example 2.29.

Exercise 3.5 In Box 3.3 the symmetric and anti-symmetric bending coefficients φ and ψ for a beam-column are given for tension as well as compression. Justify the tension formulae by use of the compression formulae and the mathematical relations

$$\cos(iz) = \cosh(z) \quad , \quad \sin(iz) = i\sinh(z)$$

between trigonometric and hyperbolic functions. (These relations follow from the definition of trigonometric and hyperbolic functions in terms of exponentials, but can also be established from the Taylor series expansions)

Exercise 3.6 Make a Taylor expansion of the beam-column stiffness matrix given in Box 3.3 including the power $(k\ell)^2$ and show that to within this approximation the stiffness matrix can be expressed as

$$
\mathsf{K} = \frac{EI}{\ell^3}
\begin{bmatrix}
12 & -6\ell & -12 & -6\ell \\
-6\ell & 4\ell^2 & 6\ell & 2\ell^2 \\
-12 & 6\ell & 12 & 6\ell \\
-6\ell & 2\ell^2 & 6\ell & 4\ell^2
\end{bmatrix}
+ \frac{N}{30\ell}
\begin{bmatrix}
36 & -3\ell & -36 & -3\ell \\
-3\ell & 4\ell^2 & 3\ell & -\ell^2 \\
-36 & 3\ell & 36 & 3\ell \\
-3\ell & -\ell^2 & 3\ell & 4\ell^2
\end{bmatrix}
$$

It is seen that in this approximate form the effect of the normal force N is contained in the second matrix, where N appears as a multiplicative factor. This matrix is often called the 'geometric stiffness matrix'. When this approximate form of the stiffness matrix is used the linearized stability problem takes the form of a linear eigenvalue problem, for which very efficient numerical techniques are available.

Exercise 3.7 Use the approximate form of the beam-column stiffness matrix derived in Exercise 3.6 to obtain approximate critical loads for the four columns shown in Box 3.2. Note, that the approximation is best, when the effective column length ℓ_e is 'large' compared with the element length. The accuracy can be improved by using two or more elements to model each column. This is a special instance of the Finite Element Method.

Exercise 3.8 In Example 3.6 it was shown that if the spring stiffness of the column in Fig. 3.14 is

$$c < 2\pi^2 \frac{EI}{\ell^3} = 2\frac{P_E}{\ell}$$

the symmetric mode would give the lowest critical load P_c. Consider the case of

$$c = \pi^2 \frac{EI}{\ell^3} = \frac{P_E}{\ell}$$

and find the critical value $(k\ell)_1$ from the transcendental equation derived in Example 3.6. Find the lowest critical load in the form $P_1/P_E = ?$.

Exercise 3.9   The shortening of a column can be evaluated via the integral

$$\Delta = \int_0^\ell (\mathrm{d}s - \mathrm{d}x) = \int_0^\ell \left(\sqrt{1 + (\mathrm{d}w/\mathrm{d}x)^2} - 1\right) \mathrm{d}x \simeq \int_0^\ell \tfrac{1}{2}(\mathrm{d}w/\mathrm{d}x)^2 \, \mathrm{d}x$$

Use the last approximation to evaluate the shortening of an Euler column with buckling mode $w(x) = w_c \sin(\pi x/\ell)$, and compare the result with the asymptotic result (3.99) for the column with fixed end (half of an Euler column).

Chapter 4

Cables

The previous two chapters dealt with beams and columns – structural elements in which bending plays a dominating role. In contrast, cables are usually assumed to be completely flexible in bending, and thus a cable must have an axial force in order to possess transverse stiffness, necessary for the support of transverse loads.

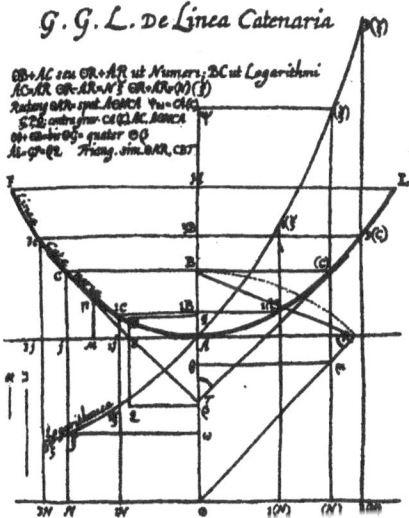

Figure 4.1: LEIBNIZ's catenary curve solution (1690).

The problem of the statics and shape of a suspended cable under the action of gravity is one of the classical problems of the mechanics of continuous bodies. The basic problem was solved nearly simultaneously by the brothers JAMES and JOHN BERNOULLI, LEIBNITZ and HUYGENS in the years 1690 and 1691, see Truesdell (1960) for an account of the early history. They discovered that a homogeneous chain, or cable, without bending stiffness would take the shape of a 'catenary', i.e. the shape of a hyperbolic cosine function.

Cables find structural application e.g. as stay cables for masts and bridges, in cable roofs and in suspension bridges, see e.g. Irvine (1981) and Gimsing (1997). While stay cables are usually freely suspended, e.g. equipped with vibration dampers, the cables in cable roofs and suspension bridges carry transverse loads.

The general static theory of a suspended inextensible cable under the action of gravity is treated in Section 4.1. The basic differential equations are derived and integrated. The general theory is illustrated by some particular solutions, and it is demonstrated that simplified approximate results may be adequate for cables with moderate sag. A simplified approximate theory for shallow cables, essentially due to Irvine (1981), is then derived in Section 4.2. In this theory the catenary profile of the cable is replaced by a parabola. Within this theory the changes of cable force and shape, introduced by the application of an additional load on the cable, are discussed in Section 4.3. Finally, Section 4.4 gives a brief discussion of cable vibrations, and identifies some of the characteristic properties.

4.1   The suspended cable

The differential equation governing the statics and shape of a suspended cable in a uniform gravitational field is derived from the sketch in Fig. 4.2. In the present context the cable is assumed to be inextensible and without bending stiffness. This implies that the cable only supports a tension force T, acting in the direction of the tangent.

Let s denote the arc-length along the cable and let $[x(s), y(s)]$ be a parametric representation of the shape of the cable in the coordinate system shown in Fig. 4.2 with horizontal x-axis and the y-axis pointing downwards. The mass per unit length of the cable is denoted m, and thus the downward force per unit cable length is mg, where g is the acceleration of gravity.

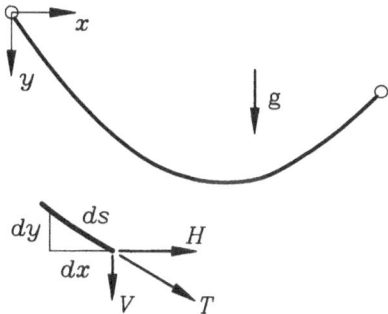

Figure 4.2: Suspended cable under the action of gravity.

The vertical force component at any section is $T(dy/ds)$, and thus vertical equilibrium requires the change of this force to balance the gravitational force,

$$\frac{\mathrm{d}}{\mathrm{d}s}\left(T\frac{\mathrm{d}y}{\mathrm{d}s}\right) = -mg \tag{4.1}$$

Note, that differentiation is with respect to the length s along the cable. The horizontal force component at any section is $T(dx/ds)$, and in the absence of any horizontal loads on the cable, the horizontal equilibrium equation becomes

$$\frac{\mathrm{d}}{\mathrm{d}s}\left(T\frac{\mathrm{d}x}{\mathrm{d}s}\right) = 0 \tag{4.2}$$

Integration of this equation gives the constant horizontal force component

$$T\frac{\mathrm{d}x}{\mathrm{d}s} = H \tag{4.3}$$

When this relation with constant H is used to eliminate T, the vertical equilibrium equation (4.1) takes the form

$$H\frac{\mathrm{d}^2y}{\mathrm{d}x^2} = -mg\frac{\mathrm{d}s}{\mathrm{d}x} \tag{4.4}$$

The arc-length increment ds is determined by

$$\mathrm{d}s^2 = \mathrm{d}x^2 + \mathrm{d}y^2 \tag{4.5}$$

and the vertical equilibrium equation has then been reduced to the following first order differential equation equation for dy/dx with x as the independent variable,

$$\frac{\mathrm{d}}{\mathrm{d}x}\left(\frac{\mathrm{d}y}{\mathrm{d}x}\right) = -\frac{mg}{H}\sqrt{1+\left(\frac{\mathrm{d}y}{\mathrm{d}x}\right)^2} \tag{4.6}$$

This differential equation is solved by introducing a variable substitution in terms of hyperbolic functions.

The hyperbolic functions $\cosh(z)$ and $\sinh(z)$ introduced in (2.111) satisfy the relation

$$\cosh^2(z) \ - \ \sinh^2(z) \ = \ 1 \tag{4.7}$$

and the differentiation rules

$$\frac{d\cosh(z)}{dz} \ = \ \sinh(z) \quad , \quad \frac{d\sinh(z)}{dz} \ = \ \cosh(z) \tag{4.8}$$

The relation (4.7) suggests the substitution

$$\frac{dy}{dx} \ = \ -\sinh[k(x - x_0)] \tag{4.9}$$

where the parameter k follows from substitution into (4.6) and use of the differentiation rule (4.9b),

$$k \ = \ \frac{mg}{H} \tag{4.10}$$

It is seen that at $x = x_0$ the slope vanishes, $dy/dx=0$, and the sign is chosen corresponding to gravity acting in the positive y-direction.

The function $y(x)$ now follows from integration,

$$y - y_0 \ = \ \int_{x_0}^{x} \frac{dy}{dx}\, dx \ = \ -\int_{x_0}^{x} \sinh[k(x - x_0)]\, dx \tag{4.11}$$

where y_0 is the coordinate corresponding to x_0. The integral is evaluated using (4.8a),

$$y - y_0 \ = \ \frac{1}{k}\Big(1 - \cosh[k(x - x_0)]\Big) \tag{4.12}$$

The arbitrary constants of the solution are the coordinates (x_0, y_0) of the point with horizontal tangent, to be determined by the boundary conditions.

The length s along the cable is determined similarly from the integral

$$s - s_0 \ = \ \int_{x_0}^{x} \frac{ds}{dx}\, dx \ = \ \int_{x_0}^{x} \sqrt{1 + \Big(\frac{dy}{dx}\Big)^2}\, dx \ = \ \int_{x_0}^{x} \cosh[k(x - x_0)]\, dx \tag{4.13}$$

where s_0 is the arc-length corresponding to the point (x_0, y_0). Integration using (4.8b) gives,

$$s - s_0 \ = \ \frac{1}{k} \sinh[k(x - x_0)] \tag{4.14}$$

The relations (4.12) and (4.14) describe the geometry of the cable.

The tension in the cable follows from (4.3) in the form

$$T = H\frac{\mathrm{d}s}{\mathrm{d}x} = H\cosh[k(x - x_0)] \tag{4.15}$$

The tension T is larger than the horizontal component H except at the point (x_0, y_0). The difference can be expressed directly in terms of the difference in height $y - y_0$ by use of (4.12) and (4.10),

$$T = H + mg(y_0 - y) \tag{4.16}$$

It follows from the definition of y_0 as the coordinate of the point with horizontal tangent, that the last term is positive.

Example 4.1

A simple illustration of the general solution to the cable equation presented above is the case of a cable of length L, suspended between two supports at the same height with distance, or span, $\ell < L$ as shown in Fig. 4.3. The solution takes a particularly simple form, if the coordinate system is located with origin at the point with horizontal tangent, but in most problems this point is not known at the beginning of the analysis, and the coordinate system is therefore centered at the left support in this example.

In the present symmetric case the parameters x_0 and y_0 of the general solution can be identified immediately as $(x_0, y_0) = (\frac{1}{2}\ell, d)$, where the span ℓ is assumed to be known together with the cable length L, while determination of the sag d is part of the problem. Also, with $s = 0$ at the left support, it follows that $s_0 = \frac{1}{2}L$.

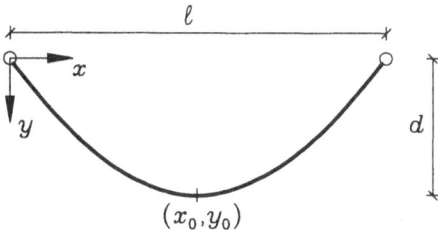

Figure 4.3: Symmetrically suspended cable with span ℓ and sag d.

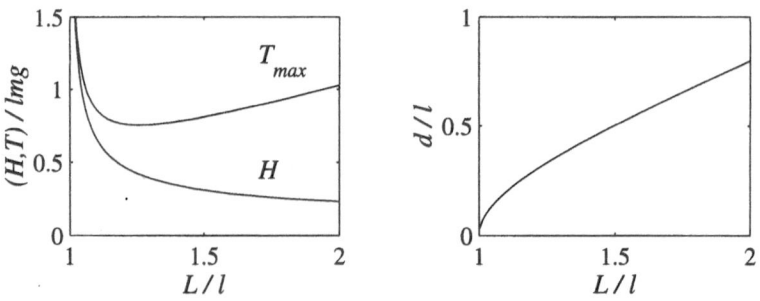

Figure 4.4: Symmetric cable, a) Forces T_{max} and H, b) Sag d.

The first step is to determine the horizontal force H. This is accomplished by use of the arc-length equation (4.14). Substitution of the left end point with $x=0$ and $s=0$ gives the equation

$$\frac{Lmg}{2H} = \sinh\left(\frac{\ell mg}{2H}\right)$$

For given L/ℓ this equation can be solved iteratively for the parameter $\ell mg/H$. A simple parametric representation of the solution can be obtained by considering $\ell mg/H$ as the independent parameter, and determining the length to span ratio as

$$\frac{L}{\ell} = \frac{\sinh\left(\dfrac{\ell mg}{2H}\right)}{\dfrac{\ell mg}{2H}}$$

This parametric representation of the cable length was used to obtain the graph in Fig. 4.4a, showing the relation between the length to span ratio L/ℓ and the horizontal force ratio ratio $H/\ell mg$.

The sag $d = y_0$ follows from substitution of the left end point $(x, y) = (0,0)$ into the relation (4.12),

$$\frac{d}{\ell} = \frac{H}{\ell mg}\left[\cosh\left(\frac{\ell mg}{2H}\right) - 1\right]$$

Also this expression is a function of the parameter $H/\ell mg$. The sag is plotted as a function of the length to span ratio in Fig. 4.4b.

The maximum cable force T_{max} is at the supports, and by substitution of the previously calculated values of H and d into (4.16),

$$T_{max} = H + dmg$$

Alternatively the ratio of T_{max} to ℓmg is expressed in terms of the parameter $H/\ell mg$ by use of (4.15),

$$\frac{T_{max}}{\ell mg} = \frac{H}{\ell mg} \cosh\left(\frac{\ell mg}{2H}\right)$$

T_{max} is also shown in Fig. 4.4a. It is seen that T_{max} attains a minimum for $L/\ell \simeq 1.25$. The reason is that for a given span ℓ a taut cable will have large tension because the cable force is nearly horizontal, while the cable force is nearly equal to half the cable weight for a very long cable. The total cable weight is $W = Lmg$, and thus the cable weight increases for increasing L/ℓ. It is seen from Fig. 4.4b that the minimum value of T_{max} corresponds to $d \simeq \frac{1}{3}\ell$. In practice this sag to span ratio is too large for most structural applications, where ratios $d/\ell \leq \frac{1}{8}$ are common.

––––––––––

The specific solution for the symmetric suspended cable is fairly simple due to the fact that the parameter x_0 appearing in the hyperbolic functions can be identified directly from symmetry. The following example illustrates the more general problem of a non-symmetrically suspended cable. In this problem the parameter x_0 must be determined as part of the solution by analysis.

Example 4.2

Figure 4.5 shows a cable that is suspended with span ℓ and with the right support h lower than the left. The coordinate system is centered at the left support, and all the parameters x_0, y_0 and s_0 are now part of the analytical solution. The figure shows the point (x_0, y_0) inside the span, but the solution procedure covers the general case.

Substitution of the left and right end points into (4.12) gives

$$\cosh(kx_0) = 1 + ky_0 \quad , \quad \cosh(kx_0') = 1 + ky_0'$$

and similarly substitution into (4.14) gives

$$\sinh(kx_0) = ks_0 \quad , \quad \sinh(kx_0') = ks_0'$$

This is a set of four non-linear equations in the unknown parameters k, x_0, y_0 and s_0.

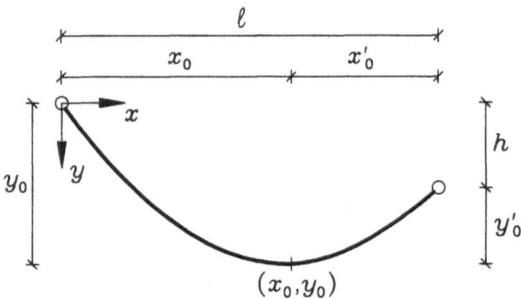

Figure 4.5: Non-symmetrically suspended cable with span ℓ.

The relations are combined by use of the following addition and subtraction formulas of hyperbolic functions

$$\cosh(z_1) - \cosh(z_2) = 2\sinh\left(\frac{z_1 + z_2}{2}\right)\sinh\left(\frac{z_1 - z_2}{2}\right)$$

$$\sinh(z_1) + \sinh(z_2) = 2\sinh\left(\frac{z_1 + z_2}{2}\right)\cosh\left(\frac{z_1 - z_2}{2}\right)$$

The parameter y_0 is eliminated by subtracting the first pair of equations and using the hyperbolic subtraction formula,

$$kh = 2\sinh\left(\tfrac{1}{2}k\ell\right)\sinh\left(\tfrac{1}{2}kb\right)$$

where $b = x_0 - x_0'$, as shown in the figure. Similarly the parameter s_0 is eliminated by adding the second pair of equations and using the hyperbolic addition formula,

$$kL = 2\sinh\left(\tfrac{1}{2}k\ell\right)\cosh\left(\tfrac{1}{2}kb\right)$$

In these equations the parameters k and b are unknown. The parameter kb is eliminated by use of the relation (4.7) between the hyperbolic sine and cosine functions. This leads to the equation

$$\left(\tfrac{1}{2}kL\right)^2 - \left(\tfrac{1}{2}kh\right)^2 = \sinh^2\left(\tfrac{1}{2}k\ell\right)$$

and by taking the square root and introducing k from (4.10), the equation takes the form

$$\frac{mg}{2H}\sqrt{L^2 - h^2} = \sinh\left(\frac{\ell\, mg}{2H}\right)$$

This equation is similar to the relation between cable length and the force ratio $\ell mg/H$ obtained for the symmetric cable in Example 4.1.

The only difference is, that in the present case $(L^2 - h^2)^{1/2}$ has to be used instead of L for the case of equal support height. The equation is solvable for

$$\sqrt{L^2 - h^2} \geq \ell$$

corresponding to

$$L \geq L_{min} = \sqrt{\ell^2 + h^2}$$

where L_{min} is the distance between the supports.

Once the parameter k has been determined, the full solution can be obtained from the relations already established. An elegant way is to add the expressions for kh and kL to obtain the relation

$$k\,(L + h) = 2\,\sinh\left(\tfrac{1}{2}k\ell\right)\exp\left(\tfrac{1}{2}kb\right)$$

The parameter kb is then obtained as

$$\frac{b\,mg}{2H} = \ln\left[\frac{(L + h)\dfrac{mg}{2H}}{\sinh\left(\dfrac{\ell\,mg}{2H}\right)}\right] = \frac{1}{2}\ln\left[\frac{L + h}{L - h}\right]$$

This determines $x_0 = \frac{1}{2}(\ell + b)$. The remaining parameters y_0 and s_0 are then given explicitly by the initial formulas.

The two examples demonstrate that although the general theory of a suspended inextensible cable is tractable, it is by no means trivial to obtain solutions to technically important problems by application of the general theory. In many structural applications of cables the sag to span ratio does not exceed 0.1. On the one hand this suggests the possibility of introducing simplifying approximations in the theory, on the other hand the use of cables with small sag leads to large cable forces as shown in Fig. 4.4a, and it may therefore be necessary to account for elastic elongation of the cable. While elastic elongation can be accomodated within the general theory, it adds new complications to the solution, see e.g. Irvine (1981) Chapter 1.

The following example illustrates the degree of accuracy that is retained in the symmetric cable problem when replacing the exact expressions with the leading term of the corresponding Taylor series expansion, asymptotically valid for small sag. It appears that the simplified algebraic results are sufficiently accurate for a sag to span ratio up to around 0.15-0.2, and the remainder of the chapter is therefore devoted to the development of an approximate theory, applicable for cables with moderate sag.

Example 4.3

The analysis in Example 4.1 of the symmetric cable problem in Fig. 4.3 consisted of the establishment of a relation between the horizontal force H and the length to span ratio L/ℓ, and expressions for the sag to span ratio d/ℓ and the maximum cable tension T_{max}. For small sag, i.e. $d/\ell \ll 1$, the horizontal force H becomes large, whereby $\ell mg/H \ll 1$. This suggests the replacement of the hyperbolic functions with a truncated Taylor series representation.

When the hyperbolic sine is replaced with a two-term Taylor expansion the length to span relation takes the form

$$\frac{Lmg}{2H} \simeq \frac{\ell mg}{2H} + \frac{1}{3!}\Big(\frac{\ell mg}{2H}\Big)^3 + \cdots$$

This reduces to

$$\frac{L}{\ell} \simeq 1 + \frac{1}{24}\Big(\frac{\ell mg}{H}\Big)^2$$

This relation is shown together with its exact counterpart in Fig. 4.6a.

An approximate expression for the sag to span ratio is obtained by replacing the hyperbolic cosine with its two-term Taylor series expansion, whereby

$$\frac{d}{\ell} \simeq \frac{H}{\ell mg}\Big[\frac{1}{2!}\Big(\frac{\ell mg}{2H}\Big)^2 + \cdots\Big]$$

corresponding to

$$\frac{d}{\ell} \simeq \frac{1}{8}\frac{\ell mg}{H}$$

This relation is shown in Fig. 4.6b together with its exact counterpart.

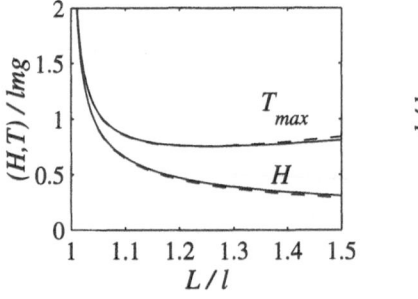 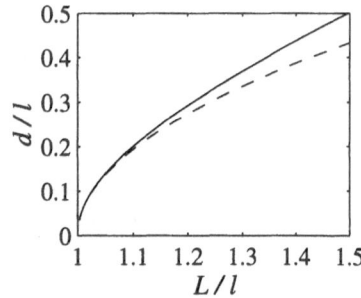

Figure 4.6: Symmetric cable. Exact (–), asymptotic (- -).

Finally the maximum cable force T_{max} was given directly in terms of H and d in Example 4.1. Substitution of the approximate expression for d leads to

$$\frac{T_{max}}{H} \simeq 1 + \frac{1}{8}\left(\frac{\ell mg}{2H}\right)^2$$

Also this relation is shown in Fig. 4.6a.

It is seen from the graphs in Fig. 4.6 that the representation of the forces H and T_{max} as function of the length to span ratio L/ℓ is excellent, while the representation of the sag is only good for $d/\ell < 0.2 - 0.25$, corresponding to $L/\ell < 1.1 - 1.15$. However, this is fully adequate for many applications.

Example 4.4

Figure 4.7 shows a marine structure moored at water depth d by a long cable of which the length $\frac{1}{2}L$ is lifted from the sea bottom. The task is to calculate the horizontal mooring force H and to calculate the lifted length of cable, when the force is increased.

As shown in Exercise 4.4 the formulas for cable length L and sag d in Example 4.1 can be combined by use of (4.7) to give,

$$\frac{1}{2}L = d\left(1 + \frac{2H}{dmg}\right)^{1/2}$$

Consider the specific case in which the initial lifted cable length is $\frac{1}{2}L_1 = 3d$. Substitution of this value into the formula gives the initial mooring force H_1 as

$$3^2 = 1 + \frac{2H_1}{dmg}$$

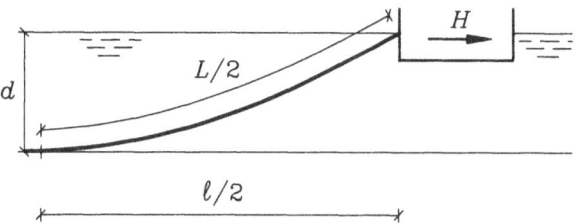

Figure 4.7: Mooring by long cable with suspended length $\frac{1}{2}L$.

whereby $H_1 = 4dmg$. If the load is changed and the mooring force is increased to $H_2 = 3H_1$, the new lifted cable length $\frac{1}{2}L_2$ follows from

$$\tfrac{1}{2}L_2 \;=\; d\left(1 + \frac{2H_2}{dmg}\right)^{1/2} \;=\; d\left(1 + 24\right)^{1/2} \;=\; 5\,d$$

Thus, the additional length, that is lifted by the increase in the mooring force is

$$\tfrac{1}{2}L_2 - \tfrac{1}{2}L_1 \;=\; 2\,d$$

The horizontal motion due to the increase of the mooring force from H_1 to H_2 is considered in Exercises 4.4 and 4.5.

4.2   Shallow cable theory

The results in Example 4.3 suggest that for shallow cables, i.e. cables where the sag to span ratio d/ℓ is small, a good approximation to the full solution can be obtained via truncating a series expansion in terms of $dy/dx \ll 1$ to only a single non-trivial term. In shallow cable theory the approximation $dy/dx \ll 1$ is introduced into the equilibrium equation, and the solutions are obtained from the approximate equation. The requirement of 'small' sag is satisfied for $d < \frac{1}{8}\ell$, corresponding to $H > \ell mg$, but in practice the theory may produce useful results even for d/ℓ up to $0.2 - 0.3$.

Suspended cable solution

The shallow cable equilibrium equation follows from (4.6),

$$\frac{\mathrm{d}^2 y}{\mathrm{d}x^2} \;=\; -\frac{mg}{H}\frac{\mathrm{d}s}{\mathrm{d}x} \;=\; -\frac{mg}{H}\sqrt{1 + \left(\frac{\mathrm{d}y}{\mathrm{d}x}\right)^2} \;\simeq\; -\frac{mg}{H}\left[1 + \frac{1}{2}\left(\frac{\mathrm{d}y}{\mathrm{d}x}\right)^2 \cdots\right] \quad (4.17)$$

by truncating the series expansion, leaving

$$\frac{\mathrm{d}^2 y}{\mathrm{d}x^2} \;=\; -\frac{mg}{H} \qquad\qquad (4.18)$$

The specific notation used in the following is shown in Fig. 4.8. As in Chapter 2 x denotes distance from the left end, while $x' = \ell - x$ denotes the

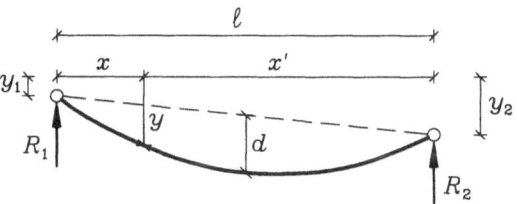

Figure 4.8: Notation for the suspended shallow cable.

distance from the right end. Similarly y_1 is the vertical coordinate of the left
support, while y_2 is the vertical coordinate of the right support.

The solution is immediately seen to be

$$y = \frac{\ell^2 mg}{2H} \frac{x}{\ell} \left(1 - \frac{x}{\ell}\right) + y_2 \frac{x}{\ell} + y_1 \left(1 - \frac{x}{\ell}\right) \qquad (4.19)$$

The sag d denotes the maximum vertical distance from the line connecting
the supports to the cable as shown in the figure. The sag follows from the
first term in (4.19) as

$$\frac{d}{\ell} = \frac{1}{8} \frac{\ell mg}{H} \qquad (4.20)$$

The solution (4.19) can then be written in the compact form

$$y = 4d \frac{x}{\ell} \frac{x'}{\ell} + y_2 \frac{x}{\ell} + y_1 \frac{x'}{\ell} \qquad (4.21)$$

This completes the solution in terms of H or d, but in many cases H or d
must be computed as part of the solution by accounting for the length of the
cable.

Within the shallow cable approximation the cable length is calculated as

$$L = \int_0^\ell \frac{ds}{dx} \, dx \simeq \int_0^\ell \left[1 + \frac{1}{2}\left(\frac{dy}{dx}\right)^2\right] dx \qquad (4.22)$$

and substitution of the solution (4.19) gives

$$L = \ell + \frac{\ell}{2}\left[\frac{1}{3}\left(\frac{\ell mg}{2H}\right)^2 + \left(\frac{y_2 - y_1}{\ell}\right)^2\right] \qquad (4.23)$$

The statics of the cable problem does not depend on y_2 and y_1 individually,
but only through the difference $\Delta y = y_2 - y_1$. The cable length can then be
written as

$$L = \ell\left[1 + \frac{1}{2}\left(\frac{\Delta y}{\ell}\right)^2\right] + \frac{\ell}{6}\left(\frac{\ell mg}{2H}\right)^2 \qquad (4.24)$$

The first term is the shallow cable approximation for the distance between the supports L_{min}, and the second term may be expressed in terms of the sag d to give

$$L = \ell\Big[1 + \frac{1}{2}\Big(\frac{\Delta y}{\ell}\Big)^2\Big] + \frac{8}{3}\frac{d^2}{\ell} \qquad (4.25)$$

The relations (4.24) and (4.25) provide an equation for H and d, respectively, when the span ℓ and cable length L are given. For equal support height, $\Delta y = 0$, these equations give the same result as the asymptotic expansion of the exact solution, derived in Example 4.3.

The cable tension T does not appear explicitly in the shallow cable theory. In effect the equilibrium equation (4.18) corresponds to a mass density m per unit *horizontal* length. Thus, the assumptions of the shallow cable problem are similar to those of the linearized beam-column theory illustrated in Fig. 3.1. Also in this case forces are conveniently represented by the horizontal component H and a vertical component V, defined by

$$V = H\frac{dy}{dx} \qquad (4.26)$$

This vertical component satisfies the equilibrium conditions specified by the equation (4.18) exactly. The corresponding consistent representation of the cable force is simply obtained as the vector sum of H and V with magnitude

$$T = \sqrt{H^2 + V^2} = H\sqrt{1 + \Big(\frac{V}{H}\Big)^2} = H\sqrt{1 + \Big(\frac{dy}{dx}\Big)^2} \qquad (4.27)$$

Expansion of the square root in this formula would introduce inconsistency in the relation between the force components H, V and T.

The vertical reactions R_1 and R_2 can be determined from the horizontal force H and the slope of the cable at the support. By differentiation of the solution in the form (4.19) the reactions are found to be

$$\begin{aligned}
R_1 &= H\frac{dy}{dx}\Big|_{x=0} = \tfrac{1}{2}\ell mg + \frac{\Delta y}{\ell}H \\
R_2 &= -H\frac{dy}{dx}\Big|_{x=\ell} = \tfrac{1}{2}\ell mg - \frac{\Delta y}{\ell}H
\end{aligned} \qquad (4.28)$$

It should be observed that these reactions could have been determined directly from the load by taking moments about the right and left support point, respectively. The last terms, $\pm H$, are due to the moment created by the horizontal force H acting at different height at the two supports. It is seen, that if one of the supports is raised, the corresponding vertical reaction is increased.

Cable elasticity and flexible supports

In the present cable theory $[x(s), y(s)]$ is a parametric representation of the suspended cable and L its length, when it supports an axial tension force of magnitude $T(s)$. If the cable is assumed to be inextensible, L is also the initial length of the cable in its unstressed state, and thus either of the equations (4.24) or (2.25) can be used to determine H and d. However, for a cable with very small sag it follows from (4.20) that the ratio $H/\ell mg$ is large, and thus the elongation of the cable due to elastic extension may become important. The effect of elastic extension of the cable can be included by expressing the length L of the suspended cable in terms of the initial length L_0, in the absence of any load.

In the suspended state the tension T depends on the location on the cable. However, the effect is only important for rather shallow cables, where the variation of T along the cable is small. Therefore, the present theory only includes the first term in the elastic correction, corresponding to replacing T with the slightly lower value H. A more detailed theory, accounting for the variation of T along the cable, corresponds to a slightly different value of the elastic properties of the cable, but does not add anything basically different.

When the elastic properties are included via the horizontal component H, the elongation of the cable is

$$\frac{L - L_0}{L_0} \simeq \frac{H}{EA} \tag{4.29}$$

where EA is the elastic stiffness of the cable, calculated as the product of the elastic modulus E and the cross-section area A. This gives the length of the suspended cable as

$$L \simeq L_0 \left(1 + \frac{H}{EA} \right) \tag{4.30}$$

Substitution of this expression into (4.24) gives the following equation for the determination of H,

$$L_0 \left(1 + \frac{H}{EA} \right) = \ell \left[1 + \frac{1}{2} \left(\frac{\Delta y}{\ell} \right)^2 \right] + \frac{\ell}{6} \left(\frac{\ell mg}{2H} \right)^2 \tag{4.31}$$

For finite elastic stiffness this constitutes a cubic equation in H. The general dependence of H on elasticity as given by (4.31) is illustrated in Fig. 4.9 for $\Delta y = 0$ and different magnitude of elastic stiffness, described by the non-dimensional parameter $EA/\ell mg$.

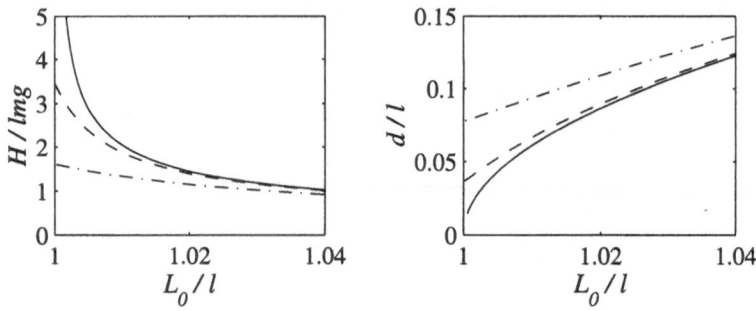

Figure 4.9: Cable force and sag, $\ell mg/EA = 0$ (–), 10^{-3} (- -), 10^{-2} (–·–).

In practice $L_0 > \ell$ and $H/EA \ll 1$. The equation is then most easily solved by iterating the expression

$$\frac{H}{\ell mg} = \frac{1}{\sqrt{24}}\left(\frac{L_0}{\ell}\left[1 + \frac{\ell mg}{EA}\frac{H}{\ell mg}\right] - \left[1 + \frac{1}{2}\left(\frac{\Delta y}{\ell}\right)^2\right]\right)^{-1/2} \qquad (4.32)$$

The iteration starts with $H = 0$ at the right side, producing the value of H for an inextensible cable.

Example 4.5

In the Great Belt suspension bridge $d/\ell \simeq 0.1$ in the main span. By (4.20) this corresponds to

$$H/\ell mg \simeq \tfrac{1}{8}\,10.0 = 1.25$$

and by (4.25)

$$L/\ell = 1 + \tfrac{8}{3}(d/\ell)^2 \simeq 1.0267$$

These values can be identified in Fig. 4.9.

In the completed configuration the mass m corresponds to the mass per length of one cable and half the deck. In the Great Belt suspension bridge $\ell mg/EA \simeq 2 \cdot 10^{-3}$, leading to a relative cable elongation of

$$\frac{L - L_0}{L_0} = \frac{\ell mg}{EA}\frac{H}{\ell mg} \simeq 2 \cdot 10^{-3} \cdot 1.25 = 2.5 \cdot 10^{-3}$$

As seen from Fig. 4.9 the effect of this elongation is small.

Box 4.1: Shallow cable equations

The equilibrium equation of a shallow cable with mass m per unit length suspended under the action of gravity g is

$$\frac{\mathrm{d}^2 y}{\mathrm{d}x^2} = -\frac{mg}{H}$$

where H is the horizontal force component. In the notation of Fig. 4.8 the freely suspended shallow cable with span ℓ takes the shape

$$y = 4d \frac{x}{\ell}\frac{x'}{\ell} + y_2 \frac{x}{\ell} + y_1 \frac{x'}{\ell}$$

with sag d determined by

$$\frac{d}{\ell} = \frac{1}{8}\frac{\ell mg}{H}$$

When the elastic cable stiffness is EA and the supports have horizontal spring stiffness k_1 and k_2, respectively, the horizontal force H is determined from the equation

$$L_0 \left[1 + \left(\frac{1}{EA} + \frac{1}{L_0 k_1} + \frac{1}{L_0 k_2}\right) H\right] = L_{min} + \frac{\ell_0}{24}\left(\frac{\ell_0 mg}{H}\right)^2$$

where L_0 is the initial cable length, ℓ_0 is the span and L_{min} is the direct distance between the supports.

Flexibility of the supports can also be included in the theory. Only a brief indication is included here. If the supports permit horizontal motion with spring constant k_1 and k_2 at the left and right support, respectively, the actual span will be shorter due the the motion of the supports. When ℓ_0 denotes the initial span, the actual span is

$$\ell = \ell_0 - \left(\frac{1}{k_1} + \frac{1}{k_2}\right) H \qquad (4.33)$$

When considering a symmetrically suspended cable $\Delta y = 0$. If furthermore the last term in (4.31) - accounting for the difference between L and ℓ - is assumed to be small, the correction can be introduced only in the first term, leading to

$$L_0 \left(1 + \frac{H}{EA}\right) = \ell_0 - \left(\frac{1}{k_1} + \frac{1}{k_2}\right) H + \frac{\ell_0}{6}\left(\frac{\ell_0 mg}{2H}\right)^2 \qquad (4.34)$$

Collecting the flexibility terms on the right then gives

$$L_0 \left[1 + \left(\frac{1}{EA} + \frac{1}{L_0 k_1} + \frac{1}{L_0 k_2}\right) H \right] = \ell_0 \left[1 + \frac{1}{24}\left(\frac{\ell_0 mg}{H}\right)^2\right] \qquad (4.35)$$

It is seen that the terms from the cable flexibility and the flexibility of the supports add up to a single flexibility term. The horizontal cable force H can be determined iteratively as in (4.31).

4.3   Static load on a cable

The shape of a cable depends on the distribution of the load. The shallow cable solution in the previous section corresponds to a uniformly distributed load of magnitude mg per unit horizontal length. The shape of the cable only depends on the magnitude magnitude of mg through the flexibility of the cable and supports. For an inextensible cable with rigid supports the shape is independent of the magnitude mg of the uniformly distributed load, as seen from (4.21) and (4.25). In contrast, the shape of the cable changes, when a load is applied locally as shown in Fig. 4.10.

Figure 4.10 shows a shallow cable of mass m per unit length with a vertical load P applied at the distance a form the left support and distance $a' = \ell - a$ from the right support. The shape of the cable is expressed in the form $y(x) + v(x)$, where $y(x)$ is the shape of the cable without the concentrated load P. Similarly, the horizontal force is written as $H + h$, where H is the horizontal force in the cable without the load P. With this notation, the shallow cable equation (4.18) takes the form

$$(H + h)\frac{d^2}{dx^2}(y + v) = -mg \quad , \qquad x < a \, , \, x > a \qquad (4.36)$$

At the force the solution must satisfy suitable continuity conditions. However, it turns out to be simpler to determine the vertical reactions directly from the load, and then integrate the solution to the left and to the right of the force independently. Due to symmetry only the left part needs to be integrated explicitly.

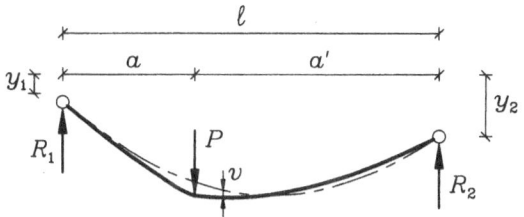

Figure 4.10: Concentrated load P on suspended cable.

The vertical reactions R_1 and R_2 are determined by taking moments about the right and left support point, respectively.

$$R_1 = \tfrac{1}{2}\ell mg + \frac{a'}{\ell}P + \frac{\Delta y}{\ell}(H+h)$$

$$R_2 = \tfrac{1}{2}\ell mg + \frac{a}{\ell}P - \frac{\Delta y}{\ell}(H+h) \tag{4.37}$$

The differential equation (4.36) is now integrated in the interval $0 \leq x < a$ to give

$$(H+h)\frac{\mathrm{d}}{\mathrm{d}x}(y+v) = -mg\,x + C_1 \qquad , \qquad x < a \tag{4.38}$$

This is recognized as the vertical force component, and the integration constant C_1 is therefore equal to the vertical force at $x = 0$, i.e. the vertical reaction R_1. Subtraction of the similar equation for y with $P = 0$ and horizontal force H gives the following differential equation for the additional displacement $v(x)$,

$$(H+h)\frac{\mathrm{d}v}{\mathrm{d}x} = P\frac{a'}{\ell} + h\frac{\Delta y}{\ell} - h\frac{\mathrm{d}y}{\mathrm{d}x} \qquad , \qquad x < a \tag{4.39}$$

The final integration with the boundary condition $v(0) = 0$ gives

$$(H+h)v = P\frac{a'}{\ell}x + h\frac{\Delta y}{\ell}x - h(y-y_1) \qquad , \qquad x \leq a \tag{4.40}$$

Substitution of $y(x)$ from (4.21) then gives the final result

$$v = \frac{P\ell}{H+h}\left[\frac{a'}{\ell}\frac{x}{\ell} - 4\frac{h}{P}\frac{\mathrm{d}}{\ell}\frac{x'}{\ell}\frac{x}{\ell}\right] \qquad , \qquad x \leq a \tag{4.41}$$

The solution for $x > a$ follows from (4.41) by interchanging x with x' and exchanging a' with a,

$$v = \frac{P\ell}{H+h}\left[\frac{a}{\ell}\frac{x'}{\ell} - 4\frac{h}{P}\frac{\mathrm{d}}{\ell}\frac{x}{\ell}\frac{x'}{\ell}\right] \qquad , \qquad x \geq a \tag{4.42}$$

The second term is quadratic and represents a decrease in curvature due to the increase in the horizontal force from H to $H + h$.

At the point $x = a$ the two parts of the solution are continuous, with the additional vertical displacement of the force

$$v_a = \frac{P\ell}{H + h} \left[1 - 4 \frac{h}{P} \frac{d}{\ell} \right] \frac{a'}{\ell} \frac{a}{\ell} \tag{4.43}$$

It should be noted, that the factor h/P with the load P in the denominator is a result of the grouping of the terms, and does not lead to a singularity for $P = 0$.

Full determination of the solution requires an equation for the additional horizontal force h. Let a point (x, y) on the cable before application of the load P move to the position $(x + u, y + v)$ after the load has been applied. The corresponding arc-length increments can then be expressed as

$$ds^2 = dx^2 + dy^2 \quad , \quad ds_*^2 = (dx + du)^2 + (dy + dv)^2 \tag{4.44}$$

The relative elongation of the line element ds can then be expressed as

$$\frac{ds_* - ds}{ds} = \sqrt{\left(\frac{dx}{ds} + \frac{du}{ds} \right)^2 + \left(\frac{dy}{ds} + \frac{dv}{ds} \right)^2} - 1$$

$$= \sqrt{1 + \left(\frac{du}{ds} \right)^2 + 2 \frac{du}{ds} \frac{dx}{ds} + \left(\frac{dv}{ds} \right)^2 + 2 \frac{dv}{ds} \frac{dy}{ds}} - 1$$

$$\simeq \frac{du}{ds} \frac{dx}{ds} + \frac{dv}{ds} \frac{dy}{ds} + \frac{1}{2} \left(\frac{dv}{ds} \right)^2 \tag{4.45}$$

where the square root has been replaced by its two-term Taylor expansion and the term $(du/ds)^2$ has been omitted, as being small compared to unity.

The extension of the cable, as given by (4.45), depends on the elastic stiffness of the cable EA and the increase of the cable force. The increase of the cable force is approximately given by $h(ds/dx)$, and thus the elastic relation of the cable is

$$\frac{ds_* - ds}{ds} = \frac{h}{EA} \frac{ds}{dx} \tag{4.46}$$

The two relations (4.45) and (4.46) are now combined to

$$\frac{h}{EA} \left(\frac{ds}{dx} \right)^3 = \frac{du}{dx} + \frac{dv}{dx} \frac{dy}{dx} + \frac{1}{2} \left(\frac{dv}{dx} \right)^2 \tag{4.47}$$

Integration of the equation over the span gives

$$\frac{h}{EA}\frac{L_e}{} = u_2 - u_1 + \int_0^\ell \frac{dv}{dx}\frac{dy}{dx}\,dx + \int_0^\ell \frac{1}{2}\left(\frac{dv}{dx}\right)^2 dx \qquad (4.48)$$

where L_e is the 'equivalent' length, defined as

$$L_e = \int_0^\ell \left(\frac{ds}{dx}\right)^3 dx = \int_0^\ell \left[1 + \frac{3}{2}\left(\frac{dy}{dx}\right)^2 + \cdots\right] dx \simeq 3L - 2\ell \quad (4.49)$$

The equivalent cable length L_e is slightly larger than the actual cable length L. For a sag to span ratio of $d/\ell = 0.1$ the cable length is given by (4.25) as $L/\ell = 1 + \frac{8}{3} \cdot 0.1^2 = 1.0267$, whereby $L_e/\ell = 1.08$.

The evaluation of the two integrals in (4.48) is rather elaborate, but is included for completeness. The reader may simply accept the results (4.56) and (4.57) and continue from there.

The integrals are evaluated via integration by parts. The first integral is reduced to

$$\int_0^\ell \frac{dv}{dx}\frac{dy}{dx}\,dx = \left[v\frac{dy}{dx}\right]_0^\ell - \frac{d^2y}{dx^2}\int_0^\ell v\,dx = \frac{mg}{H}\int_0^\ell v\,dx \qquad (4.50)$$

with $v(0) = v(\ell) = 0$, and using (4.18) to evaluate the constant second derivative.

In the second integral the integration must be carried out for the intervals $(0, a_-)$ and (a_+, ℓ) separately, because dv/dx is discontinuous at $x = a$. Integration by parts with $v(0) = v(\ell) = 0$ then gives

$$\int_0^\ell \frac{dv}{dx}\frac{dv}{dx}\,dx - v_a\left(\frac{dv}{dx}\Big|_{a_+} - \frac{dv}{dx}\Big|_{a_-}\right) - \frac{d^2v}{dx^2}\int_0^\ell v\,dx \qquad (4.51)$$

The discontinuity of dv/dx at $x = a$ follows from the fact that the cable force must hold equilibrium with the vertical load P, whereby

$$(H + h)\left(\frac{dv}{dx}\Big|_{a_+} - \frac{dv}{dx}\Big|_{a_-}\right) = -P \qquad (4.52)$$

The second derivative follows from combination of (4.36) and (4.18),

$$\frac{d^2v}{dx^2} = \frac{h}{H+h}\frac{mg}{H} \qquad (4.53)$$

Substitution of these results into (4.51) gives the second integral as

$$\int_0^\ell \left(\frac{dv}{dx}\right)^2 dx = v_a \frac{P}{H+h} - \frac{h}{H+h}\frac{mg}{H}\int_0^\ell v\,dx \qquad (4.54)$$

Thus, both of the integrals appearing in the equation (4.48) have been expressed in terms of the integral of the additional vertical displacement v.

The vertical displacement v is given by (4.41) for $x < a$ and (4.42) for $x > a$. The first term in these expressions make up a triangle with maximum at $x = a$, while the second term is a parabola with maximum at $x = \frac{1}{2}\ell$. The integral can therefore be written down directly, using the integration formulas of Appendix B,

$$\int_0^\ell v\,dx = \frac{P\ell}{H+h}\left[\frac{\ell}{2}\frac{a}{\ell}\frac{a'}{\ell} - \frac{2\ell}{3}\frac{h}{P}\frac{d}{\ell}\right] \qquad (4.55)$$

The final results for the two integrals are then obtained by substitution of this and v_a from (4.43). In terms of suitable non-dimensional parameter combinations the results are,

$$\int_0^\ell \frac{dv}{dx}\frac{dy}{dx}\,dx = \ell\frac{(\ell mg)^2}{(H+h)H}\left[\frac{1}{2}\frac{P}{\ell mg}\frac{a}{\ell}\frac{a'}{\ell} - \frac{1}{12}\frac{h}{H}\right] \qquad (4.56)$$

and

$$\int_0^\ell \left(\frac{dv}{dx}\right)^2 dx =$$

$$\ell\left(\frac{\ell mg}{H+h}\right)^2\left\{\left[\left(\frac{P}{\ell mg}\right)^2 - \frac{P}{\ell mg}\frac{h}{H}\right]\frac{a}{\ell}\frac{a'}{\ell} + \frac{1}{12}\left(\frac{h}{H}\right)^2\right\} \qquad (4.57)$$

In these relations the parameters appear in the non-dimensional combinations $P/(\ell mg)$, $H/(\ell mg)$ and h/H.

The purpose of these computations is to reduce (4.48) to an algebraic equation for the additional horizontal force h. The integrals have now been evaluated an all that remains is the representation of the support stiffness. If the supports are rigid $u_1 = u_2 = 0$. This may be considered as a limiting case of flexible supports with spring stiffness k_1 and k_2 at the left and right support, respectively. In the case of flexible supports the horizontal displacement of the supports are given by

$$u_1 = \frac{h}{k_1} \quad , \quad u_2 = -\frac{h}{k_2} \qquad (4.58)$$

where positive h implies inward motion.

Box 4.2: Concentrated load on shallow cable

The shape of a shallow cable of span ℓ with a concentrated vertical load P at distance a from the left support and $a' = \ell - a$ from the right is described as $y(x) + v(x)$, where the initial shape $y(x)$ is given in Box 4.1, and the vertical displacement $v(x)$ is

$$
v = \begin{cases} \dfrac{P\ell}{H+h}\left[\dfrac{a'}{\ell}\dfrac{x}{\ell} - 4\dfrac{h}{P}\dfrac{d}{\ell}\dfrac{x'}{\ell}\dfrac{x}{\ell}\right] \ , & x \le a \\[4mm] \dfrac{P\ell}{H+h}\left[\dfrac{a}{\ell}\dfrac{x'}{\ell} - 4\dfrac{h}{P}\dfrac{d}{\ell}\dfrac{x}{\ell}\dfrac{x'}{\ell}\right] \ , & x \ge a \end{cases}
$$

The additional horizontal force h depends on the combined stiffness of cable and supports via the stiffness parameter λ^2,

$$
\frac{1}{\lambda^2} = \left[\frac{1}{EA} + \frac{1}{L_e k_1} + \frac{1}{L_e k_2}\right]\frac{L_e}{\ell}\frac{H^3}{(\ell m g)^2}
$$

and the magnitude and location of the load through the parameter

$$
F = \left(\frac{P}{\ell m g}\right)\left(1 + \frac{P}{\ell m g}\right)\frac{a}{\ell}\frac{a'}{\ell}
$$

These two parameters determine the ratio h/H by the cubic equation

$$
\left(\frac{h}{H}\right)^3 + \left[2 + \frac{\lambda^2}{24}\right]\left(\frac{h}{H}\right)^2 + \left[1 + \frac{\lambda^2}{12}\right]\left(\frac{h}{H}\right) - \frac{\lambda^2}{2}F = 0
$$

In the limiting case of an inextensible cable with rigid supports $1/\lambda^2 = 0$, and the equation is quadratic.

Substitution of the relations (4.56)–(4.58) into (4.48) leads to the following cubic equation in h/H,

$$
\left[\frac{1}{EA} + \frac{1}{L_e k_1} + \frac{1}{L_e k_2}\right]\frac{L_e}{\ell}\frac{H^3}{(\ell m g)^2}\left(1 + \frac{h}{H}\right)^2\frac{h}{H} =
$$
$$
-\frac{1}{12}\left(\frac{h}{H}\right)^2 - \frac{1}{24}\frac{h}{H} + \frac{1}{2}\left(\frac{P}{\ell m g}\right)\left(1 + \frac{P}{\ell m g}\right)\frac{a}{\ell}\frac{a'}{\ell}
$$

(4.59)

In this equation the combined elastic flexibility of cable and supports appears in the form of a single non-dimensional parameter,

$$\frac{1}{\lambda^2} = \left[\frac{1}{EA} + \frac{1}{L_e k_1} + \frac{1}{L_e k_2}\right]\frac{L_e}{\ell}\frac{H^3}{(\ell mg)^2} \tag{4.60}$$

When the load P is applied, the cable on the two sides of the load is extended, partly due to elastic flexibility and partly due to diminished curvature. In the present notation, introduced by Irvine (1981), the parameter λ^2 represents the relative elastic stiffness. Thus, an inextensible cable with rigid supports would correspond to an infinite value of λ^2. It is important to note that the horizontal force H plays an important role in the definition of λ^2. For a taut cable $H/\ell mg$ is large, and thus λ^2 can be large, even for a stiff cable, if the sag is sufficiently small.

The load also appears in (4.59) in the form of a single non-dimensional parameter

$$F = \left(\frac{P}{\ell mg}\right)\left(1 + \frac{P}{\ell mg}\right)\frac{a\,a'}{\ell\,\ell} \tag{4.61}$$

It is seen that the load produces the largest h, when applied at the center of the cable.

In terms of the stiffness and load parameters λ^2 and F the cubic equation (4.59) takes the form

$$\left(\frac{h}{H}\right)^3 + [2 + \frac{\lambda^2}{24}]\left(\frac{h}{H}\right)^2 + \left[1 + \frac{\lambda^2}{12}\right]\left(\frac{h}{H}\right) - \frac{\lambda^2}{2}F = 0 \tag{4.62}$$

It is seen that the cubic is increasing for $h/H \geq 0$, and thus a positive load, $F > 0$, leads to a single positive root. A more detailed discussion of the roots for various load and stiffness combinations is given by Irvine (1981).

A first impression of the influence of the elastic stiffness parameter λ^2 can be obtained from the linearized solution for $h/H \ll 1$. This corresponds to $P/\ell mg \ll 1$, and when higher powers of these parameters are neglected, the solution to (4.62) is

$$\frac{h}{H} \simeq \frac{6}{1 + 12/\lambda^2}\frac{a\,a'}{\ell\,\ell}\frac{P}{\ell mg} \tag{4.63}$$

The corresponding vertical displacement of the load follows from substitution into (4.43),

$$\frac{v_a}{\ell} \simeq \frac{P}{H}\left[1 - \frac{3}{1 + 12/\lambda^2}\frac{a'\,a}{\ell\,\ell}\right]\frac{a'\,a}{\ell\,\ell} , \quad x \leq a \tag{4.64}$$

The properties of the fully non-linear solution for different values of the stiffness parameter are illustrated in the following two examples.

Example 4.6

A cable with large sag will have a small ratio $H/\ell mg$, and therefore the elastic stiffness parameter will typically be large. In the limit of infinite stiffness (4.62) reduces to the quadratic equation

$$\frac{1}{24}\left(\frac{h}{H}\right)^2 + \frac{1}{12}\left(\frac{h}{H}\right) - \frac{1}{2}F = 0$$

with solution

$$\frac{h}{H} = \sqrt{1 + 12F} - 1$$

or, after substitution of F from (4.61),

$$\frac{h}{H} = \sqrt{1 + 12\frac{a'}{\ell}\frac{a}{\ell}\left(\frac{P}{\ell mg}\right)\left(1 + \frac{P}{\ell mg}\right)} - 1$$

The vertical displacement of the force follows by substituting the ratio h/H into (4.43), conveniently written as

$$\frac{v_a}{d} = \frac{8}{1 + h/H}\left[\frac{P}{\ell mg} - \frac{1}{2}\frac{h}{H}\right]\frac{a'}{\ell}\frac{a}{\ell}$$

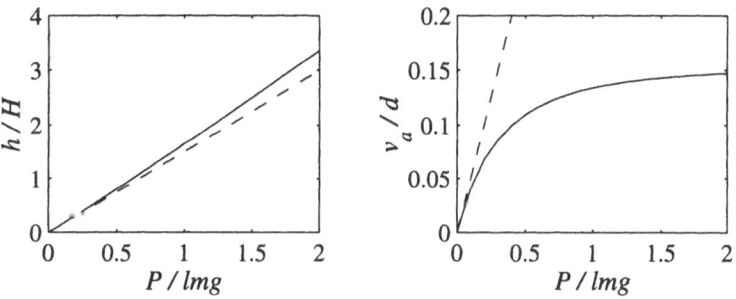

Figure 4.11: Load P at center of inextensible cable.

The solution is illustrated for a load P at the center of the span in Fig. 4.11. Also the linearized solution is indicated. It is seen that while h is estimated fairly well from the linearized theory, the load-displacement relation is highly non-linear even for for $P \simeq 0.1\ell mg$.

Example 4.7

The effect of flexibility is a combination of the elastic flexibility of cable and supports and the geometric flexibility due to the sag of the cable. For the present argument it is sufficiently accurate to assume $L_e \simeq \ell$. When the elastic flexibility is due to the cable alone, the stiffness parameter λ^2 is

$$\lambda^2 \simeq \frac{EA}{H}\left(\frac{\ell mg}{H}\right)^2 = \frac{EA}{H}\left(\frac{8d}{\ell}\right)^2$$

where the importance of the sag is seen clearly.

The effect of sag and cable flexibility is illustrated by considering a taut and a slack cable for two different values of the elastic stiffness. Let the elastic stiffness take the values already used in Fig. 4.9, namely $EA/\ell mg = 10^3$ and 10^2. Each of the cables are studied in a slack configuration with $d/\ell = 1/8$, corresponding to $H/\ell mg = 1$, and a taut configuration $d/\ell \simeq 1/36$, corresponding to $H/\ell mg \simeq 4.5$. These combinations give $\lambda^2 = 10^3$, 10^2, 10, 1 as shown in the table.

λ^2		$EA/(\ell mg)$	
		10^3	10^2
$\dfrac{H}{\ell mg}$	1	10^3	10^2
	4.65	10	1

A local load is applied at the center, the ratio h/H is calculated from (4.62), and the displacement of the force is calculated as in the previous example. The results are illustrated in Fig. 4.12, showing the force ratio h/H and the displacement ratio v_a/d.

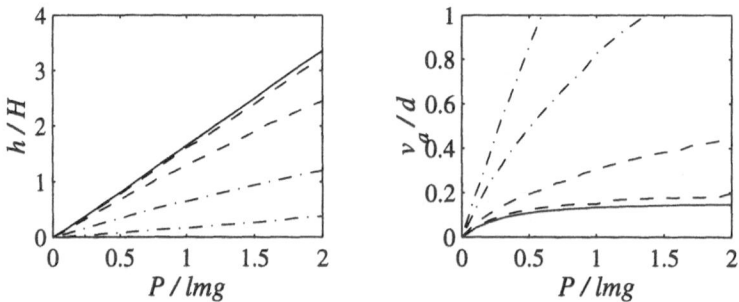

Figure 4.12: Load at center of flexible cable, 'slack' (- -), 'taut' (-·-).

It is seen from Fig. 4.12a that the force ratio h/H grows approximately linearly with the load also for the flexible cables, and the magnitude decreases with decreasing stiffness parameter λ^2 as predicted by the linearized solution (4.63). Figure 4.12b shows the displacement of the load, normalized with respect to the initial sag, and it is therefore to be expected, that the taut cable shows greater relative displacement. However, it should also be noted that the flexible cables exhibit a less non-linear load-displacement behaviour.

The methods of static cable analysis presented here can be extended in several ways. Thus, explicit solutions can be obtained for loads distributed over part of the span, see Irvine (1981), and the shallow cable results can be cast in the form of 'cable elements' with non-linear stiffness combining the effect of elasticity and cable sag. However, this is beyond the cope of this text, and this chapter concludes with a brief discussion of the cable vibration problem.

4.4 Shallow cable dynamics

In the dynamics of a suspended shallow cable the motion is predominantly in the transverse direction. This means that the inertia forces can be included in the vertical equilibrium equation as an extra term. As in the previous section the equilibrium configuration is described by the shape $y(x)$ and the horizontal force H. The additional motion $v(x, t)$ introduces a varying additional horizontal force $h(t)$. The equation of motion follows directly from (4.36) by including the inertial force $m\partial^2 v/\partial t^2$ as a load on the right side of the equation.

$$(H + h)\frac{\partial^2}{\partial x^2}(y + v) = -mg + m\frac{\partial^2 v}{\partial t^2} \tag{4.65}$$

This is a non-linear differential equation, because it contains the product $h\,\partial^2 v/\partial x^2$, but if the vibration amplitude is small this term will be of higher order and can be omitted. This leads to the linearized equation of motion

$$H\frac{\partial^2 v}{\partial x^2} - m\frac{\partial^2 v}{\partial t^2} = -mg - (H + h)\frac{d^2 y}{dx^2} \tag{4.66}$$

The dependence on the initial curvature is eliminated by use of the equilibrium equation (4.18), whereby the equation of motion reduces to

$$\frac{\partial^2 v}{\partial x^2} - \frac{m}{H}\frac{\partial^2 v}{\partial t^2} = \frac{mg}{H}\frac{h}{H} \tag{4.67}$$

This may look like an non-homogeneous differential equation, but it should be realized, that within the linearized theory the additional horizontal force $h(t)$ is proportional to the displacement $v(x,t)$. Thus, the differential equation is homogeneous, i.e. without external input. An additional equation is needed in order to determine $h(t)$ in terms of the displacement function $v(x,t)$.

The equation for $h(t)$ was obtained in the previous section in the form of a relation between the displacement increment (du, dv) and the increments dx, dy and ds of the equilibrium configuration. Within the linearized theory the relation (4.47) becomes

$$\frac{h}{EA}\left(\frac{ds}{dx}\right)^3 = \frac{du}{dx} + \frac{dv}{dx}\frac{dy}{dx} \tag{4.68}$$

where the last quadratic term has been omitted. Integration of this relation over the span gives

$$\frac{h\,L_e}{EA} = u_2 - u_1 + \int_0^\ell \frac{dv}{dx}\frac{dy}{dx}\,dx \tag{4.69}$$

where the equivalent length L_e was introduced in (4.49). Support flexibility is introduced by (4.58), and the integral is rewritten by use of (4.50). This gives the linearized cable equation in the form

$$\left[\frac{L_e}{EA} + \frac{1}{k_1} + \frac{1}{k_2}\right] h = \frac{mg}{H}\int_0^\ell v\,dx \tag{4.70}$$

It was found in the previous section that the magnitude of combined elastic stiffness and cable was conveniently expressed in terms of the non-dimensional parameter λ^2, defined in (4.60). In terms of this parameter, the linearized cable equation becomes

$$\frac{1}{\lambda^2}\frac{h}{H} = \frac{H}{\ell mg}\frac{1}{\ell^2}\int_0^\ell v\,dx \tag{4.71}$$

The left side represents the elastic flexibility and the right side 'the need for flexibility' imposed by the motion $v(x,t)$.

The two equations (4.67) and (4.71) form the basis for linearized dynamic analysis of shallow cables. They were established by Irvine & Caughy (1974). In the following the free vibrations of a shallow cable are analyzed.

Vibrations of shallow cables

The linearized equations of shallow cable dynamics (4.67) and (4.71) can describe small amplitude harmonic vibrations of shallow cables. In the harmonic vibration problem, a solution is sought in the form

$$v(x,t) = \tilde{v}(x)\cos(\omega t) \quad , \quad h(t) = \tilde{h}\cos(\omega t) \qquad (4.72)$$

where t is time, and ω is the angular frequency of the vibration. The principle of the analysis is similar to that of beam vibrations, treated in Section 2.9.

Substitution of the representation (4.72) into the equation of motion (4.67) leads the the following ordinary differential equation,

$$\frac{d^2\tilde{v}}{dx^2} + \beta^2\,\tilde{v} = \frac{mg}{H}\frac{\tilde{h}}{H} \qquad (4.73)$$

where the parameter β has been introduced by

$$\beta^2 = \frac{m\,\omega^2}{H} \qquad (4.74)$$

The differential equation (4.73) must be solved for a dynamic cable force $\tilde{h}$ that satisfies the cable equation (4.71), that is for

$$\frac{1}{\lambda^2}\frac{\tilde{h}}{H} = \frac{H}{\ell mg}\frac{1}{\ell^2}\int_0^\ell \tilde{v}\,dx \qquad (4.75)$$

It is a remarkable property of this equation, that if the integral of the displacement vanishes, there will be no change in the horizontal force, i.e. $\tilde{h} = 0$, independent of any elastic flexibility. This implies that an anti-symmetric displacement mode will correspond to $\tilde{h} = 0$. It is therefore convenient briefly to discuss this simple case first, and then revert to the general vibration modes.

Anti-symmetric modes. For anti-symmetric modes of vibration $\tilde{h} = 0$, and the differential equation (4.73) therefore simplifies to

$$\frac{d^2\tilde{v}}{dx^2} + \beta^2\,\tilde{v} = 0 \qquad (4.76)$$

This is the well known equation for vibrations of a taut string. When the supports are fixed vertically $v(0) = 0$ and $v(\ell)$, and the anti-symmetric vibration modes are

$$\tilde{v} = C_n\sin\left(\beta_n x\right) = C_n\sin\left(n\frac{2\pi}{\ell}x\right) \quad , \quad n = 1,2,\cdots \qquad (4.77)$$

with the parameters $\beta_n = n2\pi/\ell$. The corresponding angular frequencies ω_n follow from (4.74) as

$$\omega_n = \beta_n \sqrt{\frac{H}{m}} = n \frac{2\pi}{\ell} \sqrt{\frac{H}{m}} \quad , \quad n = 1, 2, \cdots \qquad (4.78)$$

This is identical to the modes and frequencies of a taut string, independent of the sag and elasticity of the (shallow) cable. The explanation is illustrated in Fig. 4.13. While the curvature is reduced at one side when the cable is lifted, the curvature is increased by a the same amount at the other side, where the cable is lowered. Thus, this mode of deformation does not involve stretching of the cable itself. Clearly, the center of the cable must move horizontally to permit this mechanism. While the inertial forces from horizontal motion were discarded in setting up the theory, the theory still contains the horizontal motion. In fact, the horizontal motion can be determined by integration of the cable relation (4.68) from $x = 0$ to any point x within the span, see Exercise 4.9.

Figure 4.13: First anti-symmetric vibration mode.

Symmetric modes. The symmetric solution to the differential equation (4.73) with boundary conditions $v(0) = 0$ and $v(\ell) = 0$ is

$$\tilde{v} = \frac{1}{\beta^2} \frac{mg}{H} \frac{\tilde{h}}{H} \left[1 - \frac{\cos[\beta(x - \frac{1}{2}\ell)]}{\cos(\frac{1}{2}\beta\ell)} \right] \qquad (4.79)$$

The solution $\tilde{v}$ is seen to be proportional to $\tilde{h}$, and substitution of $\tilde{v}$ from (4.79) into the linearized cable equation (4.75) gives the equation

$$\frac{(\beta\ell)^2}{\lambda^2} = \frac{1}{\ell} \int_0^\ell \left[1 - \frac{\cos[\beta(x - \frac{1}{2}\ell)]}{\cos(\frac{1}{2}\beta\ell)} \right] dx \qquad (4.80)$$

which does not contain $\tilde{h}$. When the integration is carried out, the equation becomes

$$\frac{(\beta\ell)^2}{\lambda^2} = 1 - \frac{1}{\frac{1}{2}\beta\ell} \tan(\frac{1}{2}\beta\ell) \qquad (4.81)$$

This equation will only be satisfied for particular values of the parameter β, and by (4.74) these values determine the vibration frequencies ω_n. The equation is conveniently rewritten in the form

$$\tan(\tfrac{1}{2}\beta\ell) = (\tfrac{1}{2}\beta\ell) - \frac{4}{\lambda^2}(\tfrac{1}{2}\beta\ell)^3 \tag{4.82}$$

An identical equation was encountered in Example 3.6, describing the critical load of a column, supported by a transverse spring at the center.

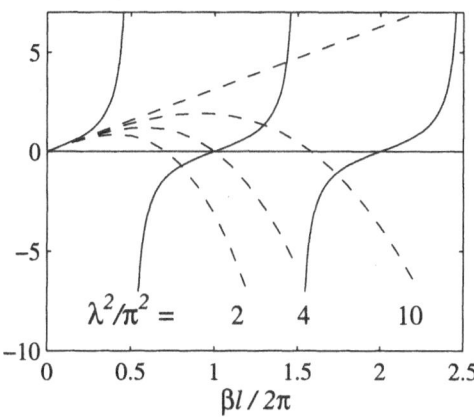

Figure 4.14: Left and right side of the cable frequncy equation (4.82).

The frequency equation (4.82) is illustrated in Fig. 4.14, showing the left and right side of the equation. For an inextensible cable with rigid supports $1/\lambda^2 = 0$, and the cubic term in the equation vanishes. This is shown as the straight line corresponding to the equation

$$\tan(\tfrac{1}{2}\beta\ell) = (\tfrac{1}{2}\beta\ell) \tag{4.83}$$

This equation was encountered and solved in Example 3.3. The smallest root was found to be $\tfrac{1}{2}\beta_1\ell = 4.493$, corresponding to the angular frequency $\omega_1 = 8.99(H/m\ell^2)^{1/2}$.

Cable flexibility introduces the negative cubic term, and with decreasing stiffness λ^2 the root $\tfrac{1}{2}\beta_1\ell$ of the equation, and thereby the angular frequency ω_1, decreases. In the limit $\lambda^2 = 0$ the point of intersection is at minus infinity, whereby $\tfrac{1}{2}\beta_1\ell = \tfrac{1}{2}\pi$, corresponding to the angular frequency $\omega_1 = \pi(H/m\ell^2)^{1/2}$. This is the familiar result for a taut string. It is seen that the flexibility of the cable may change the frequency of the lowest symmetric mode with almost a factor of three. Cable flexibility also influences the higher symmetric modes, but the effect is less pronounced.

The parameter value $\lambda^2 = 4\pi^2$ occupies a special position. For this value of the stiffness, the root of the equation is $\frac{1}{2}\beta\ell = \pi$, and the angular frequency is $\omega_1 = 2\pi(H/m\ell^2)^{1/2}$. This is precisely the same frequency as that of the first anti-symmetric mode. The reason is that, with $\beta\ell = 2\pi$ both modes correspond to one period of the corresponding trigonometric function. Thus, the wave length is equal to the span ℓ in both cases for this particular value of the stiffness.

Figure 4.15: Symmetric displacement function $\tilde{v}_1/d$. a) Flexible cable, $\lambda^2 = 2\pi^2$, b) Stiff cable, $\lambda^2 = 10\pi^2$.

The difference between the symmetric vibrations of a flexible and a stiff cable is illustrated in Figs. 4.15 and 4.16, showing the displacement function $\tilde{v}(x)$ and the corresponding mode of vibration, respectively. The stiffness parameters of the two cables are those shown in Fig 4.14, i.e. $\lambda^2 = 2\pi^2 \simeq 20$ for the flexible cable, and $\lambda^2 = 10\pi^2 \simeq 100$ for the stiff cable. The corresponding roots of the equation are $\beta_1\ell = 5.037 = 1.603\pi$ and $\beta_1\ell = 8.141 = 2.591\pi$, respectively. This implies that for the flexible cable the wavelength is larger than the span, while for the stiff cable it is smaller. The implication is shown in Fig. 4.15. It is seen that the displacement function $\tilde{v}(x)$ of the flexible cable has the same sign over the whole span and finite slope at the ends. In contrast displacement function $\tilde{v}(x)$ of the stiff cable changes sign inside the span and has inverted slope at the ends.

Figure 4.16: First symmetric cable vibration mode. a) Flexible cable, $\lambda^2 = 2\pi^2$, b) Stiff cable, $\lambda^2 = 10\pi^2$.

Box 4.3: Shallow cable vibrations

The vibration modes $\tilde{v}(x)$ of a shallow cable are determined by the equation

$$\frac{d^2\tilde{v}}{dx^2} + \beta^2\,\tilde{v} = \frac{mg}{H}\frac{\tilde{h}}{H}$$

where $\tilde{h}$ is the corresponding dynamic horizontal force, and the angular frequency ω appears through the parameter

$$\beta^2 = \frac{m\,\omega^2}{H}$$

The dynamic horizontal force is determined from the vibration mode by

$$\frac{1}{\lambda^2}\frac{\tilde{h}}{H} = \frac{H}{\ell m g}\frac{1}{\ell^2}\int_0^\ell \tilde{v}\,dx$$

where λ^2 is the combined cable and support stiffness parameter

$$\frac{1}{\lambda^2} = \Big[\frac{1}{EA} + \frac{1}{L_e k_1} + \frac{1}{L_e k_2}\Big]\frac{L_e}{\ell}\frac{H^3}{(\ell m g)^2}$$

The anti-symmetric vibration modes, with $\tilde{h}=0$, are

$$\frac{\tilde{v}}{d} = C_n \sin\Big(n\frac{2\pi}{\ell}x\Big) \quad , \quad \omega_n = n\frac{2\pi}{\ell}\sqrt{\frac{H}{m}}$$

The symmetric vibration modes are

$$\frac{\tilde{v}}{d} = \frac{8\,C_n}{(\beta_n\ell)^2}\Big[1 - \frac{\cos[\beta_n(x - \frac{1}{2}\ell)]}{\cos(\frac{1}{2}\beta_n\ell)}\Big] \quad , \quad \frac{\tilde{h}}{H} = C_n$$

with the parameter $\beta_n\ell$ determined by the equation

$$\tan(\tfrac{1}{2}\beta\ell) = (\tfrac{1}{2}\beta\ell) - \frac{4}{\lambda^2}(\tfrac{1}{2}\beta\ell)^3$$

The explanation for the behaviour of the displacement function $\tilde{v}(x)$ follows from the modes of vibration, illustrated in Fig. 4.16. In the flexible cable the motion is downward (upward) over the full span, due to the elongation (shortening) of the cable. In contrast, the stiff cable moves downward (up-ward) at the center and upward (downward) near the supports, due to the limited elongation (shortening) of the cable. The particular transition stiff-ness $\lambda = 4\pi^2$, separating the two types of motion, corresponds to unchanged direction of the cable at the supports, i.e. $d\tilde{v}/dx = 0$ at $x = 0$ and $x = \ell$. This accounts for the match in wavelength with the anti-symmetric mode.

4.5   Summary

This chapter gives a brief survey of the mechanics and analysis of ideal cables without bending stiffness. In the absence of bending stiffness the cable acts via a tension force in the direction of the tangent of the cable. In the basic cable problem the load is provided by the weight mg per unit length along the cable. This leads to a non-linear second order differential equation (4.4), which is integrated by a substitution in terms of hyperbolic functions. The solution contains the horizontal component of the cable force as a parameter. In most practical problems the cable length and the location of the supports are given, and the horizontal force component must therefore be determined from a non-linear equation as illustrated in Examples 4.1 and 4.2. A series expansion of the general solution in Example 4.2 suggests that for shallow cables a fairly good approximation can be obtained by prescribing a constant load per unit horizontal length.

The shallow cable theory, using constant load per unit horizontal length, is developed in Section 4.2, and it is demonstrated how a moderate cable flexi-bility and flexibility of the supports can be included in the theory. The basic equations of shallow cable theory are summarized in Box 4.1. The remain-der of the chapter is devoted to the study of two particular problems: the application of a local vertical static load at a point of the cable, summarized in Box 4.2, and small amplitude vibrations of shallow cables, summarized in Box 4.3. In both cases the solution is obtained by considering the displace-ment $v(x)$ and the horizontal force h, that occur in addition to the cable deflection $y(x)$ and horizontal force H of the static equilibrium configuration of the freely suspended cable.

An essential feature of the solutions to local load and vibration problems for shallow cables is the 'stiffness' of the cable. When the additional motion $v(x)$ requires extension of a part of the cable, this extension is produced partly by decreasing the sag, and partly by the elastic flexibility of the cable and the supports. Therefore the stiffness of the cable appears in the form of a non-dimensional stiffness parameter λ^2, defined in (4.60), that combines the effect of cable and support flexibility and the extensibility of the cable due to a reduction of the sag. Typically a taut cable will have a small stiffness parameter, say $\lambda^2 \leq 10$, while a slack cable will have a large value of the stiffness parameter, say $\lambda^2 \geq 50$. The effect of stiffness in the static load problem is illustrated in Example 4.7.

In cable vibrations the anti-symmetric modes are not influenced by the cable stiffness. Straightening of the cable on one side is compensated by increased curvature at the other side, as illustrated in Fig. 4.13. In contrast, symmetric cable vibrations are strongly influenced by the value of the stiffness parameter λ^2. The difference between the vibration modes of a flexible and a stiff cable is illustrated in Fig. 4.16. The flexible cable moves like a heavy rubber rope, with the whole cable moving up or down at the same time. In the case of a stiff cable the main deformation mechanism is straightening of the two halves, whereby the center is lowered and the parts near the supports are raised. This mechanism is characteristic of a suspended chain.

The presentation has concentrated on the main principles and the characteristics of the solutions. In practice cable problems will often be solved by numerical models. The present theory serves as a basis for the formulation of such models and for intelligent interpretation of their results.

4.6 Exercises

Exercise 4.1 Consider the case of a cable, suspended symmetrically with span ℓ. Figure 4.4 shows that the ratio $T_{max}/\ell mg$ has a minimum. Show that this minimum is attained for a cable with horizontal force H that satisfies the equation

$$\tanh \left(\frac{\ell mg}{2H} \right) = \frac{2H}{\ell mg}$$

Find the value of $H/\ell mg$, and determine the sag to span ratio d/ℓ.

Exercise 4.2 For an inextensible cable with small sag, $d/\ell \ll 1$, a small change ΔL in the cable length L implies a large change ΔH in the horizontal cable force H. Show by differentiation of the results for a symmetric suspended cable in Example 4.1 that

$$\frac{\Delta L/L}{\Delta H/H} \simeq \frac{H}{L}\frac{dL}{dH} = 1 - \left(\frac{\ell mg}{2H}\right)\coth\left(\frac{\ell mg}{2H}\right)$$

$$\simeq -\frac{1}{3}\left(\frac{\ell mg}{2H}\right)^2 + \frac{1}{45}\left(\frac{\ell mg}{2H}\right)^4 - \cdots$$

Consider the case in which the length $L_1 = 1.05\ell$. Calculate the horizontal cable force ratio $H_1/\ell mg$. Find the relative change in H, if the cable is replaced by one with length $L_2 = 1.01L_1$.

Exercise 4.3 Use the asymptotic formulas of Example 4.3 to find the properties of a symmetrically suspended cable, with prescribed sag ratio d/ℓ. Derive expressions for L/ℓ and $T_{max}/\ell mg$ in terms of d/ℓ, and find the values corresponding to $d/\ell = 0.05, 0.1, 0.2$.

Exercise 4.4 In Example 4.1 the following formulae were obtained for the length L and sag d of a symmetrically suspended cable,

$$\frac{Lmg}{2H} = \sinh\left(\frac{\ell mg}{2H}\right)$$

and

$$\frac{dmg}{H} + 1 = \cosh\left(\frac{\ell mg}{2H}\right)$$

Use the relation (4.7) to eliminate the dependence on the span ℓ, and obtain the formula

$$\frac{L}{2} = d\left(1 + \frac{2H}{dmg}\right)^{1/2}$$

Combine the two original formulas to obtain the following formula for the span ℓ,

$$\frac{\ell mg}{2H} = \ln\left[1 + (L + 2d)\frac{mg}{2H}\right]$$

Exercise 4.5 Consider the mooring example from Example 4.4 shown in Fig. 4.7. When the mooring force is increased from H_1 to H_2 the lifted cable length increases from $\frac{1}{2}L_1$ to $\frac{1}{2}L_2$. Use the second formula of Exercise 4.4 to determine the horizontal projections $\frac{1}{2}\ell_1$ and $\frac{1}{2}\ell_2$, and determine the horizontal motion of the marine structure, when the mooring force is increased from H_1 to H_2.

Exercise 4.6   Find the horizontal force H in a symmetrically suspended elastic cable with initial length exactly fitting within the supports, i.e. $L_0 = \ell$. Derive the sag ratio d/ℓ for such a cable, and calculate the values of $H/\ell mg$ and d/ℓ corresponding to the elasticity parameters $\ell mg/EA = 10^{-3}$, 10^{-2} illustrated in Fig. 4.9.

Exercise 4.7   Consider a cables for the main span of the Great Belt Suspension Bridge with the parameters: $\ell = 1624\,\text{m}$, $d = 165.25\,\text{m}$, $m = 3.28 \cdot 10^3\,\text{kg/m}$, $E = 2.0 \cdot 10^{11}\,\text{N/m}^2$, $A = 0.404\,\text{m}^2$. Find the horizontal force H, and the maximum cable force T_{max}. Assume rigid supports and determine the stiffness parameter λ^2. Find the displacement v_a under a concentrated load $P = 10^6\,\text{N}$ at the center of the span, representing the first deck section. $(g = 9.81\,\text{m/s}^2)$

Exercise 4.8   Use the cable data from Exercise 4.7 and determine the natural frequency $f = \omega/2\pi$ of the first anti-symmetric and the first symmetric mode of vibration.

Exercise 4.9   Integrate the linearized cable relation (4.68) between 0 and x to obtain an expression for the horizontal displacement of the form $u(x) - u_1$, where u_1 is the displacement at the left support. Use this relation to determine the horizontal displacement at the center of the span, $x = \frac{1}{2}\ell$, for the first anti-symmetric vibration mode (4.77), illustrated in Fig. 4.13. (For this argument it is sufficiently accurate to use $(ds/dx)^3 \simeq 1$.)

References

Bažant, Z.P. and Cedolin, L., *Stability of Structures*. Oxford University Press, Oxford, 1991.

Calladine, C.R., Theory of Shell Structures, Cambridge University Press, Cambridge, 1983.

Cook, R.D., Malkus, D.S. and Plesha, M.E., *Concepts and Applications of Finite Element Analysis, 3'rd Ed.*, Wiley, New York, 1989.

Dym, C.L., *Stability Theory and its Applications to Structural Mechanics*, Noordhoff, Leyden, 1994.

Felton, L.P. and Nelson, R.B., *Matrix Structural Analysis*. Wiley, New York, 1997.

Galilei, Galileo, *Two New Sciences*, Elzevir, 1638. (English translation, Dover, New York, 1914.)

Gere, J.M. and Timoshenko, S.P., *Mechanics of Materials, 4'th Ed.* PWS Publishing, Boston, 1997.

Gimsing, N.J., *Cable Supported Bridges, Concept and Design*. Wiley, Chichester, 1997.

Gradshteyn, I.S. and Ryzhik, I.M., *Table of Integrals, Series, and Products*, Academic Press, New York, 1980.

Hetenyi, M., *Beams on Elastic Foundation*, University of Michigan Press, Ann Arbor, Michigan, 1940.

Inman, D.J., *Engineering Vibration*, Prentice-Hall, Upper Saddle River, N.J., 1996.

Irvine, H.M., *Cable Structures*, MIT Press, Cambridge, Mass. 1981.

Irvine, H.M. and Caughy, T.K., The linear theory of free vibrations of a suspended cable, *Proceedings of the Royal Society, London*, Vol. A.341, pp. 299-315, 1974.

Kassimali, A., *Structural Analysis*, 2'nd Edition, PWS Publishing, Pacific Grove, 1999.

Keer, A.D., Elastic and viscoelastic foundation models, *Journal of Applied Mechanics*, Vol. 31, pp. 491-499, 1964.

Krenk, S., A general format for curved and nonhomogeneous beam elements, *Computers and Structures*, Vol. 50, pp. 449-454, 1994.

Krenk, S., Vissing-Jørgensen, C. and Thesbjerg, L., Efficient collapse analysis of offshore structures, *Computers and Structures*, Vol. 72, pp. 481-496, 1999.

Niordson, F.I., *Shell Theory*, North-Holland, Amsterdam, 1985.

Szabó, I., *Geschichte der mechanischen Principen*, Birkhäuser, Basel, 1977.

Thelanderson, S., *Konstruktionsberäkninger med Dator*, Studentlitteratur, Lund, 1984.

Thompson, J.M.T. and Hunt, G.W., *A General Theory of Elastic Stability*, Wiley, London, 1973.

Truesdell, C., *The Rational Mechanics of Flexible or Elastic Bodies 1638-1788*. Societatis Scientiarum Naturalium Helveticae, 1960.

Vallabhan, C.V.G., and Das, Y.C., Modified Vlasov model for beams on elastic foundation, *Journal of Geotechnical Engineering*, Vol. 117, pp. 956-966, 1991.

Washizu, K., *Variational Methods in Elasticity and Plasticity, 2'nd Ed.*, Pergamon Press, Oxford, 1974.

Zienkiewicz, O.C. and Taylor, R.L., *The Finite Element Method, Vols. 1 and 2*, McGraw-Hill, New York, 1989, 1991.

Appendix A. Beam load cases

Table A.1: Simply supported beam.

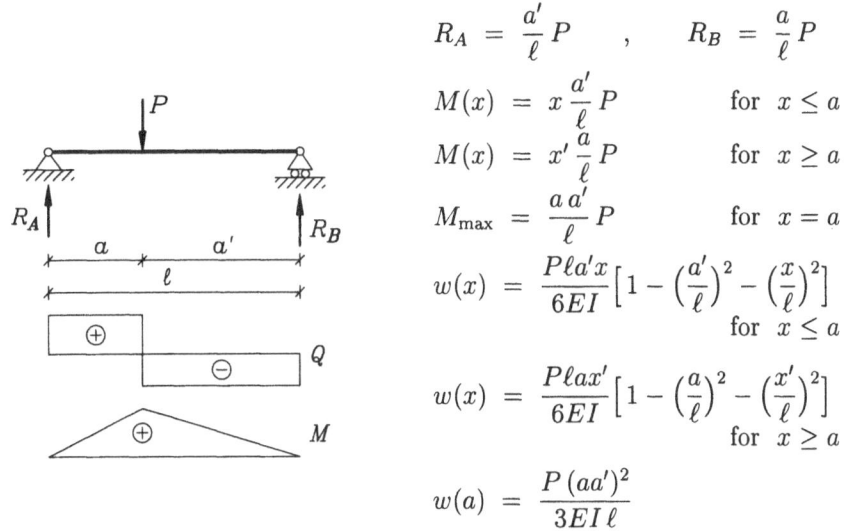

$$R_A = \frac{a'}{\ell} P \quad , \quad R_B = \frac{a}{\ell} P$$

$$M(x) = x \frac{a'}{\ell} P \qquad \text{for } x \leq a$$

$$M(x) = x' \frac{a}{\ell} P \qquad \text{for } x \geq a$$

$$M_{\max} = \frac{a\, a'}{\ell} P \qquad \text{for } x = a$$

$$w(x) = \frac{P\ell a' x}{6EI}\left[1 - \left(\frac{a'}{\ell}\right)^2 - \left(\frac{x}{\ell}\right)^2\right]$$
$$\text{for } x \leq a$$

$$w(x) = \frac{P\ell a x'}{6EI}\left[1 - \left(\frac{a}{\ell}\right)^2 - \left(\frac{x'}{\ell}\right)^2\right]$$
$$\text{for } x \geq a$$

$$w(a) = \frac{P\,(aa')^2}{3EI\,\ell}$$

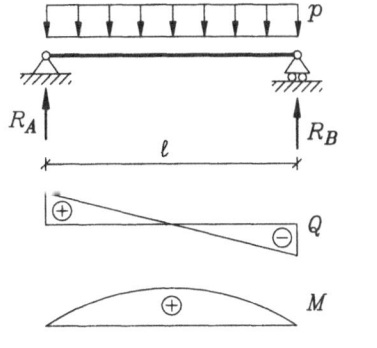

$$R_A = R_B = \tfrac{1}{2}p\ell$$

$$M(x) = \tfrac{1}{2}px(\ell - x)$$

$$M_{max} = \tfrac{1}{8}p\ell^2 \qquad \text{for } x = \tfrac{1}{2}\ell$$

$$w(x) = \frac{p\ell^0 x}{24EI}\left[1 - 2\left(\frac{x}{\ell}\right)^2 + \left(\frac{x}{\ell}\right)^3\right]$$

$$w_{max} = \frac{5}{384}\frac{p\ell^4}{EI} \qquad \text{for } x = \tfrac{1}{2}\ell$$

Table A.2: Cantilever beam.

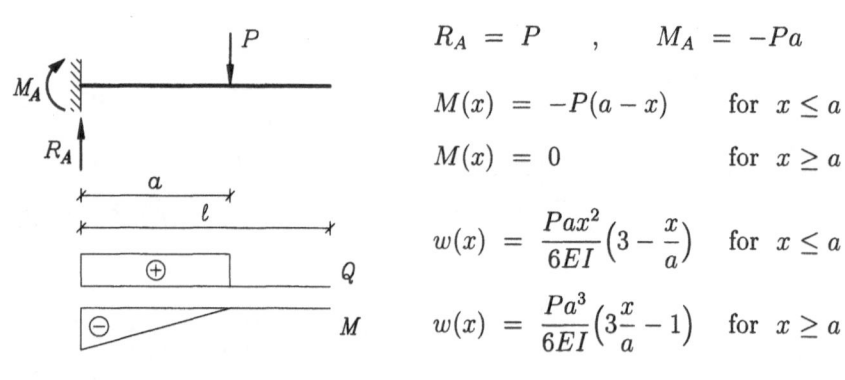

$$R_A = P \quad , \quad M_A = -Pa$$

$$M(x) = -P(a - x) \qquad \text{for } x \le a$$

$$M(x) = 0 \qquad \text{for } x \ge a$$

$$w(x) = \frac{Pax^2}{6EI}\left(3 - \frac{x}{a}\right) \quad \text{for } x \le a$$

$$w(x) = \frac{Pa^3}{6EI}\left(3\frac{x}{a} - 1\right) \quad \text{for } x \ge a$$

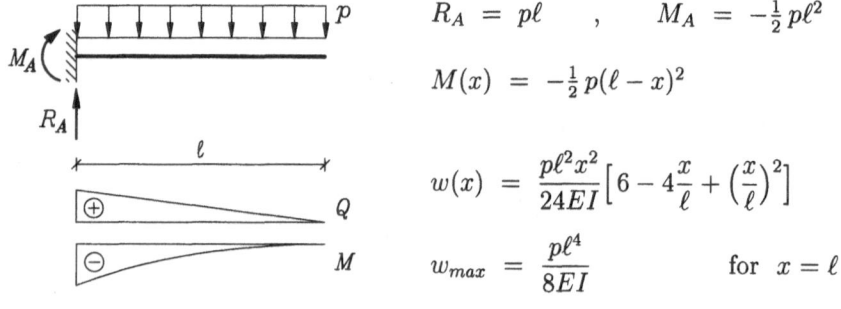

$$R_A = p\ell \quad , \quad M_A = -\tfrac{1}{2}p\ell^2$$

$$M(x) = -\tfrac{1}{2}p(\ell - x)^2$$

$$w(x) = \frac{p\ell^2 x^2}{24EI}\left[6 - 4\frac{x}{\ell} + \left(\frac{x}{\ell}\right)^2\right]$$

$$w_{max} = \frac{p\ell^4}{8EI} \qquad \text{for } x = \ell$$

Table A.3: Beam with fixed ends.

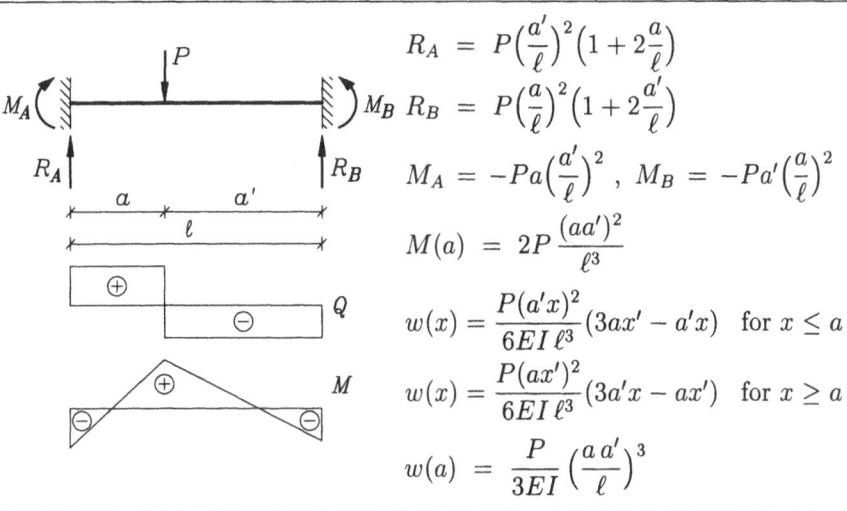

$$R_A = P\Big(\frac{a'}{\ell}\Big)^2\Big(1+2\frac{a}{\ell}\Big)$$

$$R_B = P\Big(\frac{a}{\ell}\Big)^2\Big(1+2\frac{a'}{\ell}\Big)$$

$$M_A = -Pa\Big(\frac{a'}{\ell}\Big)^2, \quad M_B = -Pa'\Big(\frac{a}{\ell}\Big)^2$$

$$M(a) = 2P\frac{(aa')^2}{\ell^3}$$

$$w(x) = \frac{P(a'x)^2}{6EI\,\ell^3}(3ax'-a'x) \quad \text{for } x \le a$$

$$w(x) = \frac{P(ax')^2}{6EI\,\ell^3}(3a'x-ax') \quad \text{for } x \ge a$$

$$w(a) = \frac{P}{3EI}\Big(\frac{a\,a'}{\ell}\Big)^3$$

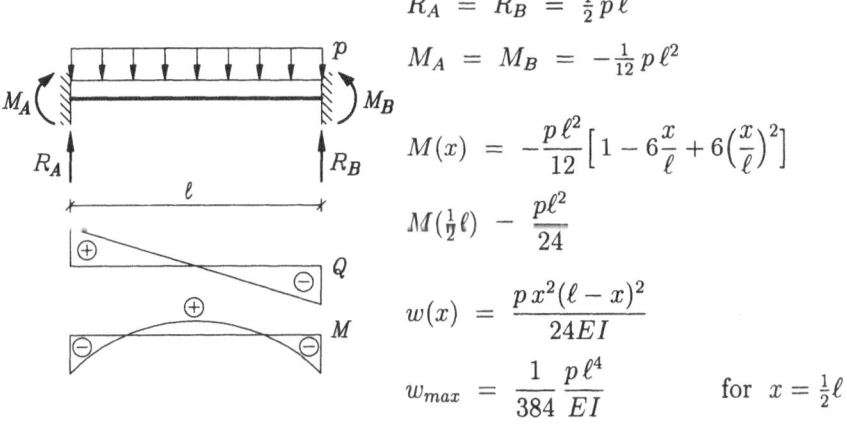

$$R_A = R_B = \tfrac{1}{2}p\ell$$

$$M_A = M_B = -\tfrac{1}{12}p\ell^2$$

$$M(x) = -\frac{p\ell^2}{12}\Big[1-6\frac{x}{\ell}+6\Big(\frac{x}{\ell}\Big)^2\Big]$$

$$M(\tfrac{1}{2}\ell) - \frac{p\ell^2}{24}$$

$$w(x) = \frac{p\,x^2(\ell-x)^2}{24EI}$$

$$w_{max} = \frac{1}{384}\frac{p\,\ell^4}{EI} \qquad \text{for } x = \tfrac{1}{2}\ell$$

Table A.4: Beam with simply supported and fixed end.

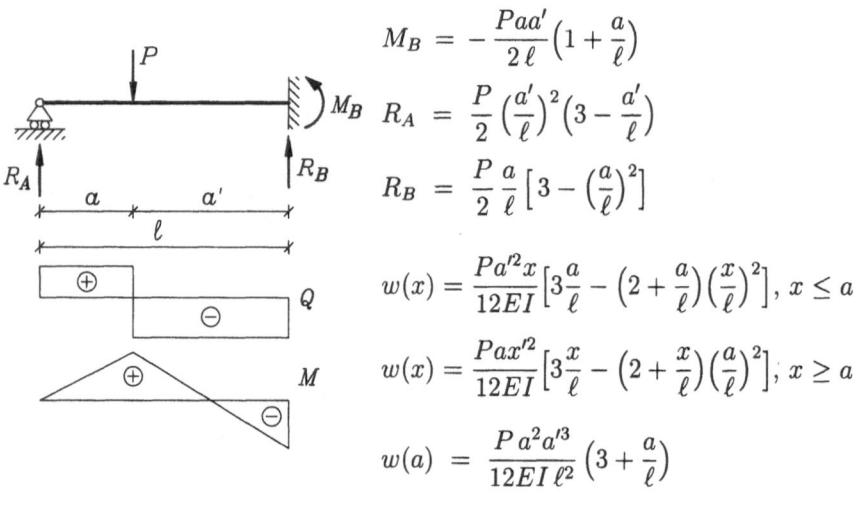

$$M_B = -\frac{Paa'}{2\ell}\left(1 + \frac{a}{\ell}\right)$$

$$R_A = \frac{P}{2}\left(\frac{a'}{\ell}\right)^2\left(3 - \frac{a'}{\ell}\right)$$

$$R_B = \frac{P}{2}\frac{a}{\ell}\left[3 - \left(\frac{a}{\ell}\right)^2\right]$$

$$w(x) = \frac{Pa'^2 x}{12EI}\left[3\frac{a}{\ell} - \left(2 + \frac{a}{\ell}\right)\left(\frac{x}{\ell}\right)^2\right], \; x \le a$$

$$w(x) = \frac{Pax^2}{12EI}\left[3\frac{x}{\ell} - \left(2 + \frac{x}{\ell}\right)\left(\frac{a}{\ell}\right)^2\right], \; x \ge a$$

$$w(a) = \frac{Pa^2a'^3}{12EI\,\ell^2}\left(3 + \frac{a}{\ell}\right)$$

$$M_B = -\tfrac{1}{8}p\ell^2$$

$$R_A = \tfrac{3}{8}p\ell \quad , \quad R_B = \tfrac{5}{8}p\ell$$

$$M(x) = \frac{p\ell^2}{8}\frac{x}{\ell}\left(3 - 4\frac{x}{\ell}\right)$$

$$M_{max} = \frac{9\,p\ell^2}{128} < |M_B| \quad \text{for } x = \tfrac{3}{8}\ell$$

$$w(x) = \frac{p\ell^3 x}{48EI}\left[1 - 3\left(\frac{x}{\ell}\right)^2 + 2\left(\frac{x}{\ell}\right)^3\right]$$

Table A.5: Beams with prescribed displacements.

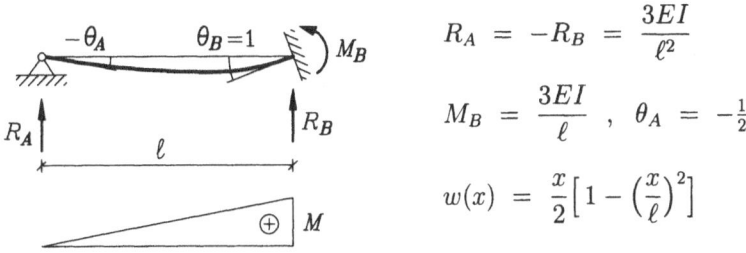

$$R_A = -R_B = \frac{3EI}{\ell^2}$$

$$M_B = \frac{3EI}{\ell} \quad , \quad \theta_A = -\frac{1}{2}$$

$$w(x) = \frac{x}{2}\left[1 - \left(\frac{x}{\ell}\right)^2\right]$$

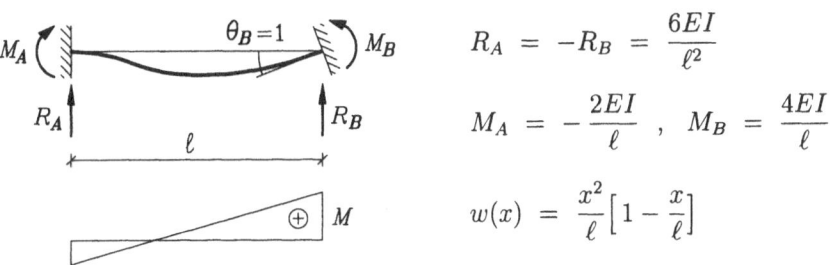

$$R_A = -R_B = \frac{6EI}{\ell^2}$$

$$M_A = -\frac{2EI}{\ell} \quad , \quad M_B = \frac{4EI}{\ell}$$

$$w(x) = \frac{x^2}{\ell}\left[1 - \frac{x}{\ell}\right]$$

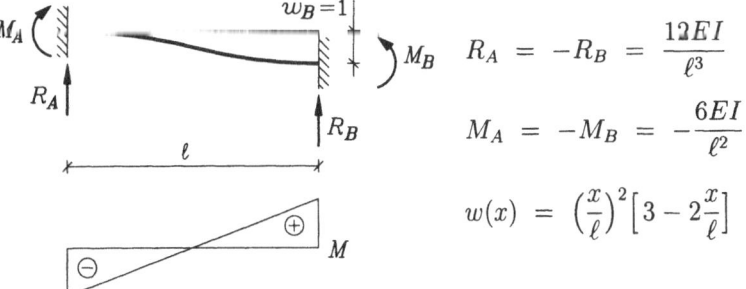

$$R_A = -R_B = \frac{12EI}{\ell^3}$$

$$M_A = -M_B = -\frac{6EI}{\ell^2}$$

$$w(x) = \left(\frac{x}{\ell}\right)^2\left[3 - 2\frac{x}{\ell}\right]$$

Appendix B. Integration formulas

$\int_0^\ell f(x)\,g(x)\,\mathrm{d}x$	$f(x)$	$g(x)$
$\dfrac{\ell}{3}\,a\,b$		
$\dfrac{\ell}{6}\,a\,b$		
$\dfrac{\ell}{6}(2ac + 2bd + ad + bc)$		
$\dfrac{\ell\,a}{3}(b + c)$		
$\dfrac{\ell\,a}{12}(3b + 5c)$		
$\dfrac{\ell\,a}{12}(b + 3c)$		

Appendix B: Integration

Index